Collins
LITTLE BOOKS

BRIDGE
SECRETS

HarperCollins Publishers
Westerhill Road
Bishopbriggs
Glasgow
G64 2QT

First Edition 2012
Second Edition 2017

Reprint 10 9 8 7 6 5 4 3 2

© HarperCollins Publishers 2017

ISBN 978-0-00-825047-8

Collins® is a registered
trademark of HarperCollins
Publishers Limited

www.collins.co.uk

A catalogue record for this
book is available from the
British Library

Author: Julian Pottage

Typeset by
Davidson Publishing Solutions

Printed and bound in China by
RR Donnelley APS Co Ltd

Acknowledgements
We would like to thank those authors
and publishers who kindly gave
permission for copyright material to be
used in the Collins Corpus. We would
also like to thank Times Newspapers Ltd
for providing valuable data.

Contents

Introduction

Bridge is a game for four players using a standard pack of 52 cards. The four players form two partnerships. The players usually sit around a square table so that nobody can see another player's cards. Each player starts with 13 cards. The action takes place in two phases, bidding and play.

The play is very similar to whist. The players each contribute one card to form a trick. If there is a trump in the trick, the player who played the highest trump wins the trick. If there are no-trumps in the trick, the highest card in the suit led wins the trick. The player who wins one trick leads to the next trick. Players have an obligation to follow suit if they can.

The important difference in the play from whist is that one player, the dummy, takes no active part in the play. That player's partner, the declarer, controls all 26 cards held by the partnership. The dummy hand is visible to all four players.

Before the play comes the bidding. The players vie to choose the trump suit (or not to have a trump suit) and to suggest how many tricks they are going to make. As there are 13 tricks, you need to make at least seven tricks in order to win more than the other side. The minimum number of tricks you can contract to make is therefore seven. This means that bidding numbers disregard the first six tricks: a bid of 'one' is a statement that you can make seven tricks, a bid of 'two' is saying you can make eight tricks and so on.

Bidding follows standard auction principles: you can only bid higher than someone else has already bid. During the bidding, the suits have the rank (starting from the lowest) of clubs, diamonds, hearts, spades, and no-trumps. (Clubs and diamonds are known as the minor suits, while the higher-ranking hearts and spades are the major suits.) Therefore, if someone else has bid 1♥, you can bid 1♠ or 1NT; if you want to bid in one of the minor suits, you must bid at least 2♣ or 2♦. If you do not wish to increase the bidding, you can pass (some say 'no bid' instead). The bidding ends when three consecutive players pass. There are also two special calls: 'double' and 'redouble'. A double (you can only double an opponent), increases the score for making or failing in a contract. Likewise, a redouble of the opponent's double further increases the scores for making a contract or the penalty for failing to make it.

You fulfil a contract if you make at least as many tricks as you have contracted to make. Thus, if the contract is 4♠, you need to score 10 or more tricks to succeed. Remember, you always add six to the number of the contract to work out the number of tricks required.

The general idea is that you increase the bidding either to outbid your opponents or in the quest for special bonuses. The most common bonus is for 'game'. To make game you need to bid and make a contract with a trick score of 100 or more. The trick scores are 20 each for clubs or diamonds and 30 each for hearts or spades. The trick scores in no-trumps are a bit more complicated. The first trick scores 40 and each subsequent trick scores 30. The minimum game contracts are therefore 3NT, 4♥, 4♠, 5♣ and 5♦.

You also get a special bonus for bidding and making a 'small slam' (12 tricks) or a 'grand slam' (13 tricks). Slams are fun, though relatively rare.

If you fail to make your contract – for example if you bid 5♦ but make only 10 tricks – penalties apply. The size of the penalty depends upon whether the opponents have doubled (and whether you have redoubled) and on whether you are 'vulnerable' or not. You become vulnerable once you have bid and made a game. The slam bonuses also depend upon vulnerability. You make a rubber when you win two games.

As I mentioned earlier, the bidding ends when three players pass. The highest bid made (possibly doubled or redoubled) becomes the contract. Whichever partner first bid the denomination of that contract becomes declarer (so if South opened 1♥ and the last bid is 4♥, South becomes declarer). This player will play two hands. The player to the left of the declarer makes the opening lead. Only after the opening leader has led does dummy appear.

If you are playing duplicate bridge, which is the norm in a bridge club, slight variations apply to the above procedures. You do not deal the cards but instead take them out of a board or wallet. You do not mix the cards in tricks but place them face down in front of you. The markings on the board, not what games you have made previously, determine which side, if any, is vulnerable.

BRIDGE
SECRETS

New suit is forcing

If partner opens in one suit and you respond in
another, this is a forcing bid. In other words, you
expect partner to bid again. The only exceptions are
if the next player overcalls, giving you another chance
to bid, or if you have previously passed. A change of
suit is forcing so that you and your partner have the
chance to exchange information about your hands.

Hand 1

♠ Q 10 9 3
♥ K 3
♦ A K J 2
♣ 6 5 4

Hand 2

♠ A J 9 3
♥ 6 3
♦ A K J 7 2
♣ 6 5

If partner opens 1♥, you respond 1♠ with Hand 1.
You do not want to have to guess that 3NT is the best
contract. Depending upon partner's shape and values,
you might belong in 4♠, 3NT, 4♥, 5♦ or even a slam.
By bidding a simple 1♠ on the first round, you have
time to find out.

If partner opens 1♥, you respond 2♦ with Hand 2. You can afford to bid your longest suit first because you have the strength to reverse into 2♠ on the second round. This will describe your hand nicely, four spades, a longer diamond suit and opening values (or close thereto). There is no way you could describe the hand in a single bid.

Eight-card fits

One of the most important objectives during the bidding is to establish whether your side has an eight-card or better fit. When you and your partner have eight or more cards between you in a suit, the suit will usually make a playable trump suit. If you have eight cards, the opponents have five, which gives you three more than they have. If one of them has three cards and the other two, which is the most likely way for the suit to divide, you will have at least two trumps left between you after their trumps have gone.

How do you know when you have an eight-card fit? If partner bids a suit, showing four, and you have four cards too, you know you have a fit. An example of that occurs when you open 1♣ and partner responds 1♥. If you have four hearts as well, you must have an eight-card or better fit. Another common example occurs when partner shows a five-card suit, with an overcall perhaps, and you have three-card support. Again, you know that your side has an eight-card or better fit.

Note that when your eight-card fit is in a minor, you may score better by playing in a higher-scoring denomination – but it is still nice to know you have the fit.

Avoid blockages

Whether you are declarer or a defender, you do not want to find that your side has winners you cannot reach.

♣ K Q J 10

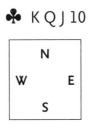

♣ A 4

Suppose these are the clubs between your two hands. If you win the first round with the king, or indeed any card other than the ace, this will block the suit. You should play the winner from the short hand to avoid this.

The next example is more subtle.

♣ K Q 5 3 2

♣ A 8 7 4

Unless all the four missing clubs are in one hand, you can make five club tricks. However, if you are not careful, you will find that the eight or seven wins the fourth round. You should cash the ace early and play the eight-seven under the king-queen to avoid that. The five is then the highest remaining card.

No-trumps and flat hands

When you do not have a long suit or any very short suits, the best spot may well be a no-trump contract. You need only nine tricks for game in no-trumps, which is one reason why a no-trump contract can be attractive.

Most no-trump bids 'limit' your hand, showing its strength within a narrowly defined range, which helps partner enormously in judging the right contract.

♠ 8 5 2
♥ A K 7
♥ Q 10 4
♣ K J 8 4

If you are playing a 12-14 1NT opening, you should open 1NT. Do not worry about the weak spades. If instead you are playing a 15-17 1NT opening, you open 1♣ and rebid 1NT after any one-level change-of-suit response.

After your no-trump bid, partner will know your side's combined strength to within a point or two. Partner will also be confident that you have tolerance for any suit.

Higher suit first

If you have two suits of equal length that you intend to bid – usually two five-card suits or sometimes two six-card suits – you should bid the higher suit first. This will keep the bidding lower.

You	Partner
1♠	1NT
2♥	

By bidding the spades first, partner has the option to leave you in 2♥ or put you back to 2♠. If instead you opened 1♥ and rebid 2♠, partner would be unable to put you back to hearts without going to the three level.

It is the same story if you are responder.

You	Partner
	1♣
1♥	1NT
2♦	

By bidding the higher suit first, you maintain the option of playing in either suit at an economical level.

A useful thing to remember follows from this: if you bid the suits in reverse order (i.e. lower first), it normally means that your first suit is longer.

Count trumps

Ideally, you want to count all the suits. Counting the trump suit is especially important.

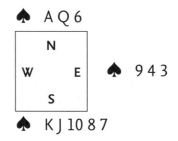

```
              ♠  A Q 6
            ┌─────────┐
            │    N    │
♠  5 2      │ W     E │      ♠  9 4 3
            │    S    │
            └─────────┘
              ♠  K J 10 8 7
```

There is an easy way to count the trumps and a hard way. Let us try the hard way first. You cash dummy's ace of spades (trumps) and all follow. That is four gone. You can still see six between your two hands, so there are three out (13–4–6=3). You then cash the king and all follow again. That is four more trumps gone, eight in all. So, with four left between your hands, there is just one out (13–8–4=1).

Now we will see the easy way. With five trumps missing, the possible breaks are 5-0, 4-1 and 3-2.

So, if someone shows out on the first round, they are 5-0 and it will take five rounds to draw them – yikes! If someone shows out on the second round, they are 4-1 and it will take four rounds to draw them. Finally, as would happen here, if all follow to two rounds, you know they are 3-2 and it will take three rounds to draw them.

Note that if the opponents ruff something, you will need to allow for that. Life is more complicated then!

Count your losers

When you are declarer in a suit contract, it is good idea to count your losers as soon as dummy appears.

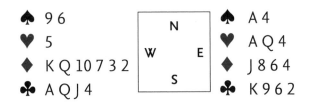

Suppose you are in 5♦ and North leads the ♠K. You count two losers, one in diamonds and one in spades. There are no heart losers because you have the ace facing a singleton and none in clubs either. In a five-level contract, you can afford two losers and so there is no need for heroics. You simply win the spade, knock out the ♦A, regain the lead, draw any missing trumps and claim.

Now suppose you have reached the more ambitious slam contract of 6♦. You still count your losers and again find two. You cannot afford two losers in a slam. You must win the spade, come to hand with a club and finesse the ♥Q. If this works, you will be able to discard a spade on the ♥A and reduce your loser count to one. If the finesse loses, you will lose a heart as well as a spade and a diamond – three losers in all – but that is a risk worth taking in the slam.

By counting your losers, you knew whether you had to risk taking the heart finesse.

Count your points

How do you value a bridge hand? You assign a value to the high cards, the cards most likely to win tricks. The usual method, sometimes referred to as Milton, is this:

ace	4 points		queen	2 points
king	3 points		jack	1 point

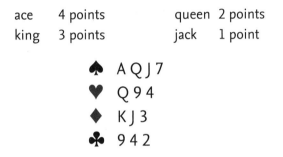

♠ A Q J 7
♥ Q 9 4
♦ K J 3
♣ 9 4 2

This hand contains 13 points, 7 in spades, 2 in hearts and 4 in diamonds.

As a partnership, the number of points you need between the two hands for various contracts is roughly as follows:

3NT (game)	25 points
6NT (small slam)	33 points
7NT (grand slam)	37 points

This page gives you the building blocks. When you are in the habit of counting points for high cards, you can also adjust for distribution and intermediate cards.

Count your winners

When you are declarer in a no-trump contract, it is a good idea to count your winners as soon as dummy appears.

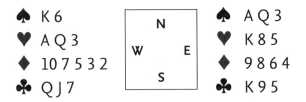

♠ K 6
♥ A Q 3
♦ 10 7 5 3 2
♣ Q J 7

♠ A Q 3
♥ K 8 5
♦ 9 8 6 4
♣ K 9 5

Let us suppose that you have bid to 3NT. North leads the ♠J. You should count the immediate winners – three in spades and three in hearts. Your contract calls for nine tricks, so you need to generate three more. The club suit cannot possibly give you three more tricks, which means you need to play on diamonds. If the diamonds divide 2-2, you will be able to set up three long diamonds, which will give you the tricks you need.

Now suppose that you have stopped in 2NT. You again count six winners. Taking six from your target of eight leaves two more to find. This is easy – by driving out the ♣A, you can guarantee to set up the two tricks you need.

With practice, you will find that counting winners is a good idea in all sorts of situations, suit or no-trump, declarer or defender. You just have to start somewhere!

Draw trumps

When you choose a trump suit, you tend to do so because you expect to have more trumps than your opponents and because you might want to do some ruffing. However, while the opposing trumps remain out, they may pose a threat to your winners.

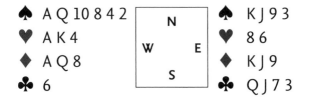

♠ A Q 10 8 4 2
♥ A K 4
♦ A Q 8
♣ 6

♠ K J 9 3
♥ 8 6
♦ K J 9
♣ Q J 7 3

You are in 6♠. North, who opened 3♥, leads the ♥Q.

You can see two possible losers, a heart and a club. You intend to avoid the heart loser by ruffing the third round in dummy. Take care not to rush into doing so. The bidding forewarns you that North holds seven hearts. If you try to cash two top hearts too early, South will surely ruff. You should draw trumps, in three rounds if necessary. Then you can go for the heart ruff and cash the winning diamonds safe in the knowledge that the opponents cannot ruff.

Tales abound of people who had to sleep rough on the Embankment in London because they failed to draw trumps. Although that fate will probably not befall you, on many hands this is still a good tip to follow!

Follow with the lowest of a sequence

As a defender, when partner leads a suit, you generally play high. What do you do if your highest cards are of equal value (say the jack and ten)? You should play the lower card. This makes the position clearer to partner.

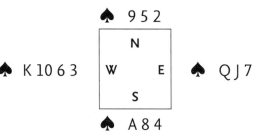

♠ 9 5 2

♠ K 10 6 3

♠ Q J 7

♠ A 8 4

Suppose partner leads the three. If your jack draws the ace, it is obvious you have the queen because declarer would prefer to take the trick with the queen rather than the ace.

You should also follow with the lower or lowest of touching cards when you are finessing against dummy.

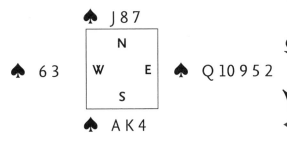

When partner leads the six and dummy plays low, you play the nine rather than the ten.

Fourth highest of your longest and strongest

Especially against no-trump contracts, it is a good idea to lead from a long suit. Declarer will be unable to ruff any long cards you set up. By leading a low card, you hope to retain an entry to your long cards.

Let us say the opponents bid 1NT-3NT. What do you lead from the two hands below?

Hand 1
♠ Q 10 7 4 2
♥ 9 5 3
♦ A K
♣ 8 5 3

Hand 2
♠ 7 4
♥ K 9 7 5 3 2
♦ J 10 5 2
♣ K

With Hand 1, you lead the ♠4, your fourth highest spade. Your top diamonds should allow you to regain the lead and continue the suit. Length takes precedence over strength.

With Hand 2, you lead the ♥5, your fourth highest heart.

Give preference

If partner bids two suits, you should usually give preference between them. To say you prefer the first suit, you go back to that suit, with a jump if your values justify doing so. To say you prefer the second suit, you can pass (if the sequence is not forcing) or raise.

On the hands below, your bidding starts 1♥-1♠-2♣.

Hand 1	Hand 2
♠ A Q 7 5 3	♠ A Q 7 5 3
♥ 8 4 2	♥ 4
♦ 7 4	♦ 10 6 4 2
♣ J 8 3	♣ J 8 3

On Hand 1, you put partner back to 2♥. Although your lengths are equal, partner will often have longer hearts than clubs, making hearts the better choice of trump suit.

On Hand 2, you prefer clubs and, with not many more values than you have already shown, you pass.

Keep a poker face

When you are playing bridge, you should aim to avoid making any gestures, mannerisms or remarks about how things are going. It is partly a matter of not giving the game away to the opponents and partly a matter of ethics, not giving unauthorized information to your partner.

If the opponents decide to play in a suit in which you have five trumps, you do not beam from ear to ear, rub your hands with glee and double in a voice of thunder!

If partner leads a suit in which you hold only small cards when you are dying for the lead of some other suit, you do not shake your head, glare or say 'tut-tut'.

If declarer drops your singleton king, you do not fling the card onto the table.

Of course, between hands, it is a slightly different matter. Then it is fine to exchange pleasantries with your partner or the opponents. You may also say 'thank you partner' when dummy appears – but if you do that then try to do so in a uniform tone whether or not dummy is as expected.

Knock out stoppers early

In a no-trump contract, you should usually hold back your top winners, preferring instead to play on the suits in which the opponents have the highest cards.

♠ K Q 6
♥ A 8 4 3
♦ K J 6
♣ Q 7 3

♠ A 4
♥ K 7 2
♦ Q 8 5 2
♣ K J 6 5

You are in 3NT and North leads the ♠J.

What happens if you begin by cashing your five certain winners in the major suits? When you lose the lead, the opponents will have a spade or two and some hearts to cash – you will go down.

The sensible approach is to knock out the opposing stoppers, the minor-suit aces, while you still have control of the major suits. This way nothing can stop you from making at least nine tricks – three spades, two hearts, two diamonds and two clubs. You might even make an overtrick or two if you find one of the minor suits 3-3 or a doubleton ace in the right place.

Lead from ace-king

The opening lead is one of the most important but also one of the most difficult plays on a deal. You usually want to set up or cash winners for your side without helping declarer too much – it can be a delicate juggling act. Leading a top card from an ace-king against a suit contract gives you the best of both worlds. You cannot set up declarer's king because you have it yourself, which makes the lead reasonably safe. Moreover, unless declarer is void, you will hold the lead, giving you all sorts of options after you have seen dummy and received a signal from partner.

The standard lead from the ace-king and others is the ace. Partner will encourage if holding the queen or (in a suit contract) a doubleton.

Against a no-trump contract, the position is slightly different. You would not normally lead from only A-K-x, preferring a long suit instead. If you have a five-card or longer suit (e.g. A-K-x-x-x) but no sure outside entry, you lead fourth highest instead of a top card. With A-K-x-x, you lead the ace, just as you would against a suit contract.

Lead partner's suit

Partners just love it when you lead their suit – well most do anyway! By leading partner's suit, you are likely to be leading up to strength, which is a good thing to do. Leading partner's suit saves you guessing what else to lead – another advantage!

Some auctions are particularly good ones for following this rule. If partner has overcalled, bid a suit strongly, or opened the suit in third seat, the suit is likely to be a good one.

Assuming you have decided to lead partner's suit, most people play that you lead the same card in the suit as you would in an unbid suit – top of a sequence, fourth highest, top of a doubleton and so on.

You can draw a useful inference if partner unexpectedly does not lead the suit you bid. If the opponents end up in spades after you have bid hearts but partner leads a club, you can guess that the club is a singleton – or possibly that partner has the ace or a void of hearts.

Lead towards strength

When you can, you should lead towards strength
rather than away from it. This allows you to capture
opposing high cards with higher ones of your own
and to save your other high cards for later if they play
their winners.

♥ A Q

```
        N
♥ K 10 6 4 2  W   E   ♥ J 9 8 3
        S
```

♥ 7 5

If you cash the ace, you make just one heart trick no
matter who holds the king. Much better is to lead
from the South hand, playing the queen unless the
king appears. This way you make two tricks whenever
West holds the king.

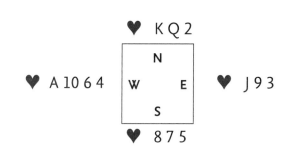

♥ K Q 2

N
W E
S

♥ A 10 6 4   ♥ J 9 3

♥ 8 7 5

If you lead from dummy, you make only one heart trick. Better is to lead from the South hand, twice if need be, using the king and queen when West does not play the ace.

The rule applies to the defenders too – though then you cannot be sure what strength you are leading toward.

Low-level doubles are for takeout

When the bidding is at a low level, it is unlikely that you can pick up a useful penalty by doubling the opponents for penalties. A much more useful use of a double is to say that you wish to compete but you have no obvious bid. In all of the auctions below, West's double is for takeout, asking partner to bid some other suit.

West	North	East	South
			1♥
Double			

West	North	East	South
	1♥	Pass	2♥
Double			

West	North	East	South
1♣	1♥	Pass	2♥
Double			

West	North	East	South
	1♦	Pass	1♥
Double			

On the first three auctions, West should have something in the three suits other than hearts. On the fourth auction, West does not need diamond support because an opponent has bid the suit.

Lead up to weakness

If you are defending and dummy is on your right, you should lead suits in which dummy is weak. Leading up to weakness is still a good idea if dummy is on your left, though then it may be harder to tell whether declarer is weak. You are hoping that by leading up to dummy's weakness you are leading up to partner's strength.

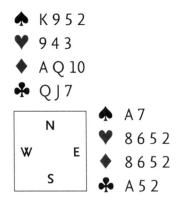

♠ K 9 5 2
♥ 9 4 3
♦ A Q 10
♣ Q J 7

♠ A 7
♥ 8 6 5 2
♦ 8 6 5 2
♣ A 5 2

You are East defending South's contract of 4♠. West leads a club, which you win with the ace.

A heart switch, up to the weakness in dummy, is much more likely to be productive than a diamond, up to its strength. There are all sorts of heart holdings partner might have, A-Q-x-x or K-J-x-x for example, that would benefit from having you lead the suit. By contrast, a diamond switch will achieve nothing: dummy's diamond honours stand ready to decapitate any high diamonds partner has.

Majors are important

The major suits, hearts and spades, are important ones to have and to bid. There are two good reasons for this. In a competitive auction, you can outbid the opponents at the same level if your side has the higher-ranking suit. In a constructive auction, you need only ten tricks for game in a major compared with eleven tricks for game in a minor. In each case, the effect is the same – you have one fewer trick to make. Since many contracts are a struggle, having to make one fewer trick is a highly desirable objective.

The desire to find a major-suit fit when you have one has a big part to play in bidding theory. If you have a major, you will often bid it, even when the alternative is raising your partner's (minor) suit. If partner opens one of a suit and you hold a major that you can show at the one level, you always bid that rather than bidding 1NT. If the opponents open and you are thinking of making a takeout double, whether you have support for the unbid major(s) can be a guiding factor in whether to do so.

Open to the rule of 20

How good a hand do you need to open the bidding? A count of your high-card (Milton) points gives you a good idea. Most hands with 12 points are worth opening. This is not the whole story. Your shape plays an important part. The usual rule 'the rule of 20' or 'Bergen rule' is to add the length of your two longest suits to your high-card points. If the answer comes to 20 or more, your hand warrants a one-level opening bid.

Hand 1
- ♠ A 8 5 2
- ♥ K 10 4 2
- ♦ K J 9
- ♣ 4 2

Hand 2
- ♠ 8
- ♥ K Q 10 4 2
- ♦ A J 9 4 2
- ♣ 4 2

Hand 1 has 11 points but only 8 cards in the two longest suits, total 19 – not enough to open.

Hand 2 has 10 points and 10 cards in the two longest suits, total 20 – enough to open (1♥).

Open your longest suit

During the bidding, a key objective is to select the best possible suit as trumps. In general, you want to end up with as many trumps as possible. You should open your longest suit (i) because by doing so partner knows it is your longest suit and (ii) if the bidding should become highly competitive or partner is very weak, you might not get another chance to show a suit.

♠ 9 7 5 4 3 2
♥ A K Q 10
♦ K 3
♣ 3

With this hand, you open 1♠, not 1♥. It is irrelevant that your hearts or so much stronger than your spades. What counts is the length rather than the strength. If the opponents play high clubs forcing you to ruff, you want to ruff with those little spades, not with the high hearts.

You also bid your longest suit first when you are making an overcall and often when you are responding too.

When it comes to selecting a trump suit, it really is a case of the more the merrier.

Overcall with a good suit

Overcalls differ from opening bids. Once the other side has opened the bidding, you are less likely to buy the contract. Moreover, if you catch partner with the wrong hand, it is more likely that they will be able to double and defeat you. If you do not buy the contract, half the time partner will be on lead and be likely to lead your suit – you want that if you have a good suit but perhaps not if your suit is poor.

Hand 1	Hand 2
♠ A K J 10 7	♠ Q 7 5 3 2
♥ J 6	♥ J 6
♦ 8 4 2	♦ K 8 4
♣ 7 4 2	♣ K 7 4

With Hand 1, you might reasonably overcall 1♠ if your right-hand opponent opens one of any other suit. With Hand 2, by contrast, you should pass. With Hand 2, you do not really want a spade lead – and nor do you want partner to think that bidding high in spades is a good idea.

Respond with 6 points up

If partner opens the bidding with one of a suit, you should keep the bidding open if you have 6 points or more (and on some hands with 5 points). Remember, partner may have 19 points or so, which means your combined total could be about 25 – if it is, you may make game. Another reason for responding is that partner's opening suit might not be the best fit for your side. A further reason is that you make life more difficult for the opponents if you bid rather than pass.

Hand 1
- ♠ 8 5
- ♥ A 8 5 4
- ♦ Q 9 4
- ♣ 10 6 3 2

Hand 2
- ♠ A J 6 3
- ♥ 7 3
- ♦ J 8 5 3 2
- ♣ 6 3

With Hand 1, you respond 1♥ if partner opens 1♣ or 1♦. You raise 1♥ to 2♥ or respond 1NT to 1♠.

With Hand 2, you respond 1♠ to 1♣, 1♦ or 1♥. You raise 1♠ to 2♠.

A simple raise of partner's suit or a 1NT response is a limited bid, showing a minimum responding hand (6-9 points). A change of suit, while consistent with a minimum hand, is a wide-ranging bid.

Note that if partner opens 1NT, you need not respond with these values. If you are weak and flat, you pass 1NT.

Return partner's suit

When partner leads a suit, especially the opening lead, it is often a good idea to return it. There are two good reasons for this. Firstly, partner must have had a reason to lead the suit in the first place. Secondly, if you open up a new suit, you risk giving away a trick in that new suit.

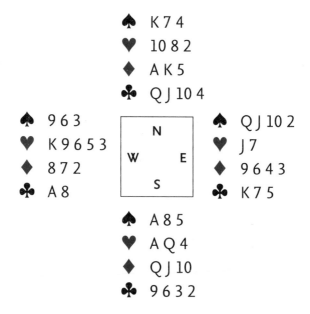

```
                  ♠ K 7 4
                  ♥ 10 8 2
                  ♦ A K 5
                  ♣ Q J 10 4

  ♠ 9 6 3          ┌─────────┐          ♠ Q J 10 2
  ♥ K 9 6 5 3      │    N    │          ♥ J 7
  ♦ 8 7 2          │  W   E  │          ♦ 9 6 4 3
  ♣ A 8            │    S    │          ♣ K 7 5
                   └─────────┘
                  ♠ A 8 5
                  ♥ A Q 4
                  ♦ Q J 10
                  ♣ 9 6 3 2
```

South plays in 3NT. West leads the ♥5. Declarer wins your ♥J with the ♥Q and plays a club. West ducks and you win with the ♣K.

Despite your attractive sequence in spades, you should return a heart. You place partner with the ♣A (or declarer would have finessed), so setting up partner's hearts should prove fruitful. Your side makes three hearts and two clubs.

Ruff in the short hand

As declarer, you should look to take ruffs in the short
trump hand. This way you generate extra tricks.
Ruffing in the long trump hand does not usually
create extra tricks because the long trumps would
be winners anyway.

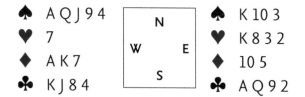

♠ A Q J 9 4
♥ 7
♦ A K 7
♣ K J 8 4

♠ K 10 3
♥ K 8 3 2
♦ 10 5
♣ A Q 9 2

You are in 6♠. North leads the ♥Q followed by the
♥J. Having ruffed this you should play to ruff the
third round of diamonds in dummy. This is much
better than trying to ruff any more hearts in hand,
which would cause you to lose control if trumps
break 4-1.

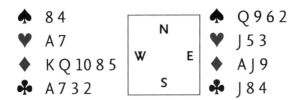

```
♠  8 4              N           ♠  Q 9 6 2
♥  A 7                          ♥  J 5 3
♦  K Q 10 8 5    W     E        ♦  A J 9
♣  A 7 3 2          S           ♣  J 8 4
```

You are in 2♦ and get a heart lead to your ace.
Because the clubs might not be 3-3, you should
immediately play on the suit. You can ruff the
fourth round in dummy to give you your eighth
trick.

Signal attitude on partner's lead

This does not mean smiling, nodding your head or frowning! You say whether you like partner's lead by the card you play. The standard signal is to play a high spot card to say you like the lead and wish to encourage its continuation. You play a low card, indeed your lowest card, if you wish to discourage a continuation.

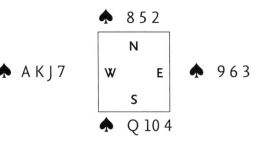

♠ 8 5 2

♠ A K J 7

♠ 9 6 3

♠ Q 10 4

If West leads the ace, you as East play the three to suggest a switch. So long as the next spade lead comes from your side of the table, South does not make the queen.

If, instead, you held Q-9-3, you would be happy to see the suit continued. You would then play the nine as a come-on, encouraging signal.

The word HELD, standing for High Encouraging, Low Discouraging, might help you to remember the signal.

Note that a signal is just that – a suggestion rather than a command. You can only indicate what looks right based on the cards you see. Partner may know better.

Sort your cards

Players the world over sort their cards into suits and into rank within suits. You will have noticed that, in this book (and any other books or magazines you might have seen), that the suits appear in the same order.

Compare these two hands:

They are in fact the same hand – but the second layout is so much easier to follow! You can see at once both the shape and the strength of the hand.

At the table, rather than following the rank of the suits, it is usual to sort the suits red-black-red-black (or black-red-black-red). This reduces the risk that you will accidentally get a heart in with your diamonds or vice versa.

Note that when you display dummy you should continue with the red-black-red-black alternation. The exception comes only when you are void in the same colour suit as the trump suit (e.g. void in clubs when spades are trumps) because putting trumps on your right takes precedence.

Stayman after 1NT

If partner opens 1NT and you have four cards in one or both majors, you will have a 4-4 fit if partner has four cards in the same suit. To find out you bid 2♣. This is the popular Stayman convention. The replies are:

- 2♦ No four-card major
- 2♥ Four hearts
- 2♠ Four spades but not four hearts

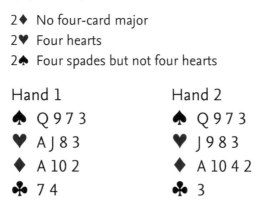

Hand 1
- ♠ Q 9 7 3
- ♥ A J 8 3
- ♦ A 10 2
- ♣ 7 4

Hand 2
- ♠ Q 9 7 3
- ♥ J 9 8 3
- ♦ A 10 4 2
- ♣ 3

After partner opens 1NT, you bid 2♣, Stayman, with either of these hands. With Hand 1, you intend to rebid in no-trumps if partner denies a four-card major (by bidding 2♦) or to raise if partner shows a major. With Hand 2, you intend to pass whatever partner does.

Note that you need to be ready for any reply from opener. If you are too weak to invite game, you must hold both majors to use Stayman. If you are stronger, one four-card major will suffice (because you can convert to no-trumps if partner denies a major). However, do not use Stayman if you have a 4333 shape – a no-trump contract will probably produce as many tricks when you lack a ruffing value.

Support partner's suit

Supporting partner's suit when you have support is one of the most important rules of bidding. One of the key aims in the bidding is to find a playable trump suit – when you find one you need to tell partner the good news.

Supporting partner's suit is especially important when the suit is a major because it is almost certain that you will want to play in that suit.

♠ A Q 7 4 2
♥ K 9 4
♦ Q J 6 2
♣ 6

You open 1♠. If partner responds 2♦, you raise to 3♦, showing the support. If partner responds 2♥, again you raise to 3♥. This is fine because the 2♥ response shows a five-card or longer heart suit.

The general rule is that if partner has shown five cards in a suit you can raise with three and that if partner has shown four cards in a suit you can raise with four.

If partner has made a bid that is likely to show a five-card suit (but not guaranteed it), it is usually fine to make a single raise with three-card support. You do not however make a jump raise with only three-card support if partner might have only four cards in the suit.

Takeout doubles show the unbid suits

If an opponent opens and you have support for several suits, it may be right to double. The double is for takeout, asking partner to bid a long or relatively long suit.

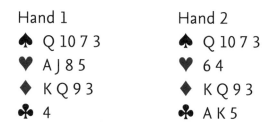

Hand 1
- ♠ Q 10 7 3
- ♥ A J 8 5
- ♦ K Q 9 3
- ♣ 4

Hand 2
- ♠ Q 10 7 3
- ♥ 6 4
- ♦ K Q 9 3
- ♣ A K 5

With Hand 1, you have a perfect double if the hand on your right opens 1♣. You have support for all the other suits. If, however, the suit opened against you is something else, you have to pass. You cannot afford to double and risk hearing 2♣ from partner.

Likewise, with Hand 2, you have a good hand for a takeout double if an opponent opens 1♥. You have support for two of the unbid suits and tolerance for the third. If instead, they open some other suit, you would have to pass. You do not want to double and hear partner leap about in hearts.

With a 4441 or 4432 hand, a double is right when you have opening values and an opponent opens in your short suit. In the same scenario, with a 5431 hand, double is best if the five-card suit is a minor and the four-card suit is a major. Otherwise, it is usually better to bid the five-card suit.

Weak takeout

If partner opens 1NT, it is not a command to play there. The thing is that by showing a balanced hand, partner is showing support or at least tolerance for any suit that you might bid. You do not remove 1NT into a four-card suit of course: often you will end up in a 4-3 'fit', which might not produce any more tricks than 1NT but which will put you at a higher level. If you have a five-card or longer major, you are likely to do better playing in that that rather than leaving partner to stew in 1NT.

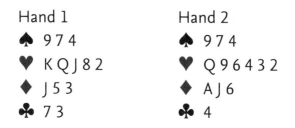

Hand 1
- ♠ 9 7 4
- ♥ K Q J 8 2
- ♦ J 5 3
- ♣ 7 3

Hand 2
- ♠ 9 7 4
- ♥ Q 9 6 4 3 2
- ♦ A J 6
- ♣ 4

Hand 1 may produce very few tricks as dummy in 1NT. Unless your partner holds the ♥A, an astute defender will hold up, killing your long suit. Bid 2♥ or, if playing transfers (as explained on page 90), 2♦.

With Hand 2, the club suit may be a weakness in 1NT. The heart suit may also be hard to set up and run as you have only one sure entry. Bid 2♥ or, if you play transfers, 2♦.

Note that you cannot take out to 2♣. A bid of 2♣ is Stayman, asking for four-card majors. If you play transfers, you cannot take out to 2♦ either.

Think at trick one

As soon as dummy appears, all the players, especially declarer should think before playing any cards. It is so often important to form a plan of how to play, something you cannot really do until dummy is on display.

♠ A 7 4   N   ♠ Q J 2
♥ A K 8 4   W   E   ♥ 9 2
♦ A K 8 4   S   ♦ 7 6 2
♣ J 4   ♣ K Q 10 9 2

You are West in 3NT and North leads the ♠10. If you do not think, you may play an honour from dummy. If the ♠K does not appear and someone holds up the ♣A, you will not reach all those lovely clubs. You need dummy's spade honours as an entry – obvious if you think about it.

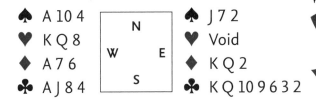

♠ A 10 4　　　　　　♠ J 7 2
♥ K Q 8　　 N　　　♥ Void
♦ A 7 6　 W　E　　♦ K Q 2
♣ A J 8 4　　S　　　♣ K Q 10 9 6 3 2

You are in 6♣ and North leads the ♥J. If you do not
think, you may ruff in dummy and subsequently lose
two spade tricks. Think, however, and you throw a
spade from dummy, losing a heart but no spades.

Watch what has gone

Watching and keeping track of the cards is crucial. You know that every deal starts with 52 cards, 13 in each player's hand. If you watch the cards that have gone, you know which remain. The more cards and tricks that have gone, the easier it is to deduce what cards a player has left.

A particular thing to watch out for is for someone to show out of a suit. Then you know that the other unseen hand has all the remaining cards in the suit.

Most players instinctively know which high cards have gone. On some deals, the spot cards are important too.

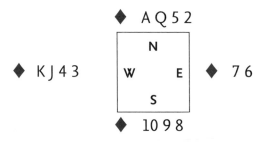

Suppose you lead the ten, covered by the jack and queen. You later lead the nine, which holds. You continue with the eight, covered by the king and ace. You will notice that East shows out on the third round. Have you also noticed that East has played the seven and six on the first two rounds, leaving North's five as a winner for the fourth round?

Top of a sequence

When you are making the opening lead, a good choice is usually a sequence of touching honours, e.g. K-Q-J. You lead the top card, which tells partner that you have the next lower honour. A lead from a three-card sequence combines safety with attack, which makes it a good choice. If the third card in the sequence is not quite touching, Q-J-9 for example, the lead is still a good one. Sequence leads are good against both suit and no-trump contracts.

Hand 1
- ♠ 8 5 3
- ♥ A 7 3
- ♦ K Q J 8
- ♣ 7 4 2

Hand 2
- ♠ 8 5 3
- ♥ A 7 3
- ♦ Q J 9 7
- ♣ 7 4 2

With Hand 1, you lead the ♦K after many auctions, such as 1NT-3NT, 1♥-3♥-4♥ or 4♠-4NT-5♦-6♠.

With Hand 2, you lead the ♦Q after many auctions, including those listed above.

Stoppers for no-trumps

If the opponents bid a suit, you need a stopper in the suit to make a no-trump bid. You expect the opponents to lead the suit and so need to make sure that they cannot run the suit against you. The same applies if your side has bid three suits – you need a stopper in the fourth suit because the opponents are most likely to lead that suit.

♠ A Q 9 4
♥ K 4
♦ 7 3 2
♣ Q 10 7 3

Suppose partner opens 1♥, you respond 1♠ and partner rebids 2♦. In this case, a 2NT rebid describes your hand well, about 11 points and a stopper in the unbid club suit. You would also bid 2NT if partner opens one of a red suit and the next hand overcalls 1♠.

Now suppose partner opens 1♥ and they overcall 2♦. With these diamonds, bidding 2NT would be foolish. A negative double is better.

Likewise, if partner opens 1♥, you respond 1♠ and partner rebids 2♣, you cannot rebid 2NT with no cover in the unbid diamond suit. A raise to 3♣ would be correct.

1NT rebids and 1NT openings

If you open one of a suit and rebid 1NT after partner has made a one-over-one response, this shows a balanced hand (usually). What strength does it show?

The answer depends upon the range of your 1NT opening. If you play a weak no-trump, 12-14, like many players do in the UK, the 1NT rebid shows 15-17.

If you play strong no-trump opening, 15-17, like many players in North America and continental Europe, the 1NT rebid shows 12-14.

♠   K 9 6
♥   Q 8 4
♦   K 10
♣   A J 8 4 3

You open 1NT if you play a weak no-trump, or open 1♣ and rebid 1NT if you play a strong no-trump, to show 12-14 points.

The ranges for the 1NT opening and 1NT rebid have to be different or you would have no sensible rebid on flat hands out of range for the 1NT opening.

At one time, many used 15-16 as the range for a 1NT rebid. This is inefficient and inadvisable. You do not want to have to jump to 2NT with 17 and to 3NT with 19.

1NT response

If partner opens the bidding with a suit higher ranking than your suit but you have too few values to respond at the two level, what do you do? You respond 1NT. This keeps the bidding alive and gives partner a chance to show other features – extra values or a second suit, for instance. You will not always have a balanced hand.

Hand 1	Hand 2	Hand 3
♠ 6 4	♠ 6	♠ 6
♥ Q 9 7	♥ Q 9 7 3	♥ Q 9 7 6 3 2
♦ 9 8 6 2	♦ K 8 6 2	♦ K 8 6
♣ A 8 4 2	♣ A 8 4 2	♣ Q 8 4

On all of these hands, you respond 1NT to 1♠. With Hand 3, you plan to rebid 2♥ if you get the chance.

Doubling artificial bids

If an opponent makes an artificial bid, such as a transfer, a cue bid, or reply to a 4NT enquiry, you can double to show that you have values in the suit you double (usually length as well). Normally you do so because you want the suit led. It is generally safe to double a suit that an opponent has bid but not shown because it is unlikely that the bidding will end with your double.

West	North	East	South
	2NT	Pass	3♦ (transfer
Double			to hearts)

West	North	East	South
	2♣	Pass	2♦ (negative
Double			or waiting)

West	North	East	South
	1♠	Pass	4♠
Pass	4NT	Pass	5♦ (one ace)
Double			

On all three auctions, West's double shows strength in diamonds and suggests that East leads a diamond against the eventual contract. In the first two, it shows length in the suit as well because the opponents could be heading for 3NT or hold length in diamonds themselves.

New suit at the three level is forcing

In uncontested auctions, bidding a new suit at the three level is normally forcing, i.e. it tells partner to continue bidding. This is useful because it saves you from having to jump to show a good hand. Having some bids as forcing leaves you room to explore.

West	North	East	South
1♠	Pass	2♥	Pass
3♣			

West	North	East	South
		1♥	Pass
1♠	Pass	2♥	Pass
3♣			

West	North	East	South
1♠	Pass	2♠	Pass
3♣			

In all three auctions, West's 3♣ is forcing. In the first two, it is natural and game-forcing. In the third auction, 3♣ is a try for game in spades and so not game-forcing.

West would need a minimum of about 15 points on the first auction, 13 points (opening values) on the second and 17 points (five and a half losers) on the third.

A reverse shows strength and shape

A 'reverse', i.e. bidding a lower ranking suit before a higher ranking suit in such a way that partner cannot go back to bid two of your first suit, is a descriptive bid. It shows that your first suit is longer and that you have the values to justify raising the bidding level.

West	North	East	South
1♦	Pass	1♠	Pass
2♥			

West has shown at least five diamonds, at least four hearts (but not equal length in the suits) and upwards of 16 points or so.

West	North	East	South
		1♦	Pass
2♣	Pass	2♦	Pass
2♠			

West has shown at least five clubs, at least four spades (but not equal length in the suits) and upwards of 11 points or so.

If you have a minimum hand or you have equal length in the suits, you should not reverse. With say five hearts, four spades and 12 points, you simply open 1♥ and rebid 2♥ if partner responds 2♣ or 2♦.

Add your dummy points

If partner bids a suit and you have support for the suit, you can add points for your short suits. This is because ruffs in the short trump hand (yours) will generate extra tricks. The number of points you should add depends upon the length of your support.

If you know of an eight-card fit, you count distributional points on the following scale:

doubleton	1 point
singleton	2 points
void	3 points

If you know of a nine-card or better fit, you count dummy points for your distribution. These are more generous because the more trumps you have the more ruffs you are likely to make. The scale is as follows:

doubleton	1 point
singleton	3 points
void	5 points

1NT overcall

Whether you play a weak or strong 1NT opening, a 1NT overcall shows strong no-trump values (15-17 or perhaps 15-18). It is too risky to come in with 12-14 when you already know that one opponent has opening values. For one thing, responder will be in a good position to judge when the opening side has the balance of power (and double you). For another, the chance of game your way has gone down. It is also important to hold a stopper or two in opener's suit – opener's partner is likely to lead the suit.

Hand 1
- ♠ K J 7 4
- ♥ A Q 10
- ♦ 5 2
- ♣ A Q 9 3

Hand 2
- ♠ A J
- ♥ K 9 4
- ♦ A Q 8 5 2
- ♣ K 8 3

With Hand 1, you overcall 1NT if the hand on your right opens 1♣, 1♥ or 1♠. If, however, the opening is 1♦, you make a takeout double – you have no diamond stopper and are happy for partner to bid one of the other suits.

With Hand 2, you overcall 1NT if the hand on your right opens one of any suit. A 5332 shape is a balanced hand. Bidding 1NT is much more descriptive than overcalling in diamonds. Besides, your best chance of game is in 3NT, not 5♦.

2♣ opening

You do not want to open a massive hand with a non-forcing opening at the one level. If your hand is so strong that you can envisage game opposite next to nothing from partner, it may be right to open 2♣. This is an artificial bid and says nothing about your holding in the club suit. Your partner must keep the bidding open, usually by responding 2♦, which is also an artificial bid.

Hand 1
♠ A J
♥ A K J 8 4
♦ A K Q 9 3
♣ 2

Hand 2
♠ A Q 5
♥ A K 4
♦ A Q 6 3
♣ K J 10

With Hand 1, you open 2♣, planning to rebid 2♥ over a 2♦ negative and later to show the diamonds. Partner must keep the bidding open.

With Hand 2, you open 2♣, planning to rebid 2NT over 2♦. That is the one sequence allowing you to stop short of game after a 2♣ opening.

With a balanced hand, you need 23 high-card points to open 2♣. With an unbalanced hand and great playing strength, you can shade the high-card requirement slightly.

Beware long-hand ruffing

Having a long or relatively long trump suit enables you to stop the opponents from running their long suits and allows you to draw their trumps. Unless you have trumps to spare, do not take unnecessary ruffs in the long trump hand.

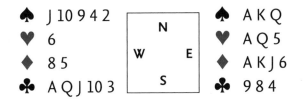

♠ J 10 9 4 2		♠ A K Q
♥ 6	N	♥ A Q 5
♦ 8 5	W E	♦ A K J 6
♣ A Q J 10 3	S	♣ 9 8 4

You play in 6♠. North leads the ♥J. You win in dummy and cash the ♠A-K. North shows out on the second round and you play a third. If you ruff a heart (or play the ♦A-K and ruff a diamond) to come to hand to draw the last trump, you will then be out of trumps and in trouble if the club finesse loses. Instead, come to hand with the ♣A.

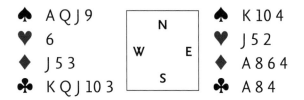

♠ A Q J 9
♥ 6
♦ J 5 3
♣ K Q J 10 3

♠ K 10 4
♥ J 5 2
♦ A 8 6 4
♣ A 8 4

You judge well to play in 4♠. North leads top hearts.
If you ruff, you will be down to three trumps and be
unable to draw trumps on a normal 4-2 break. You
should discard diamonds on the second and third
rounds of hearts.

Be a nice partner

Bridge is a game to enjoy. Part of the enjoyment of the game comes from having a partner. Unfortunately, partners make mistakes – just as you do. Most of them do not want to know when they have made them or when you think they have made them. They certainly do not want you to call them an idiot or similar. Believe me – they do not make mistakes on purpose – even if it seems that way at times!

If you have a bad board, it is usually best just to forget about it, at any rate until the end of the session. Some pairs have one disaster after another because they have a heated exchange after the first one. They are concentrating on trying to win the argument rather than focusing on the next hand. Arguing with your partner at the table is a recipe for bad results – and losing partners.

Most of the time there is no point discussing a hand immediately because the chance the same one will come up again during the session is very slim.

If something fundamental comes up – for example, partner discards the nine of spades and your spade switch is not what partner wanted – you can raise the matter. Just be nice about it – perhaps taking the blame: 'Sorry partner, we should have decided what discards we are playing. Do you want to play McKenney?'

Slam on a finesse

In general, you want to be in a small slam only if it has at least a 50% chance of success. Sometimes you cannot find out all you need to know. There may be one card partner could hold that would make a big difference. If you decide that the best you can hope for is that the slam will need a finesse, you do not want to bid it. By contrast, if you decide that at worst it will be on a finesse you should bid it.

Suppose the bidding starts as follows:

West	North	East	South
1♠	Pass	3♠	Pass
4NT	Pass	5♥ *	Pass
?			

* Two key cards (here aces) but no ♠Q

Hand 1
- ♠ K Q J 5 3
- ♥ K J 10 4
- ♦ A K Q
- ♣ 5

Hand 2
- ♠ K J 9 5 3
- ♥ K J 10 4
- ♦ A K Q
- ♣ K

You know an ace is missing. However, with Hand 1, if partner has ♥A-x or the ♥Q, 6♠ will be a very good contract. Even if not, you expect to make the slam if a finesse against the ♥Q succeeds – so bid it. With Hand 2, prospects are less good; you will need to pick the trump suit and avoid a slow heart loser – so sign off in 5♠.

Use their suit as forcing

With rare exceptions hardly worth knowing about, a bid of the opposing suit is forcing. A partner who bids a suit that one of the opponents has bid does not want to play there! A bid of the enemy suit, a 'cue bid', can have a variety of meanings depending upon the context. The common theme is that it is forcing.

West	North	East	South
			1♦
1♠	Pass	2♦	

This cue bid of 2♦ usually shows fair values and three-card spade support.

West	North	East	South
1♦	1♠	2♣	Pass
2♦	Pass	2♠	

This cue bid of 2♠ asks for a spade stopper but may be just a general forcing hand.

West	North	East	South
	1♣	2♣	Pass

Most play this cue bid of 2♣ to show a two-suited hand, normally the majors.

Covering honours

It is often right to cover declarer's or dummy's honours – but only when you might promote a lesser honour.

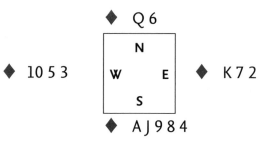

East should cover the queen with the king to promote West's ten to a third-round winner.

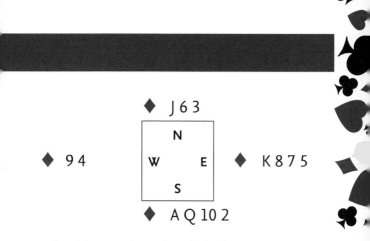

                    ♦ J 6 3

           ┌─────────────┐
           │   N   │
   ♦ 9 4   │ W   E │   ♦ K 8 7 5
           │   S   │
           └─────────────┘

                    ♦ A Q 10 2

East should cover the jack with the king to promote the eight into a fourth-round winner.

                    ♦ Q J 10 9 3

           ┌─────────────┐
           │   N   │
   ♦ 2   │ W   E │   ♦ K 8 7 5
           │   S   │
           └─────────────┘

                    ♦ A 6 4

East should not cover this time – there is nothing to promote. The king can live to win the fourth round.

Covering touching honours

When declarer leads a touching honour from either hand, you usually cover the last of the touching honours.

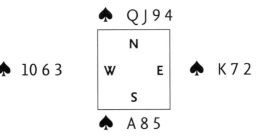

♠ Q J 9 4

♠ 10 6 3   N   ♠ K 7 2
            W   E
                 S

♠ A 8 5

If you cover the first lead of the queen, declarer wins with the ace and finesses the nine on the way back. Best is to duck the first honour but cover the second.

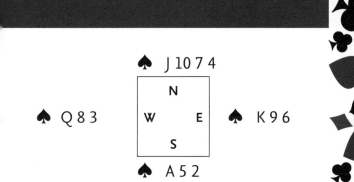

♠ J 10 7 4

♠ Q 8 3   N W E S   ♠ K 9 6

♠ A 5 2

Do not cover the jack first time. If you do, declarer takes the ace and plays up to the ten, losing only one trick.

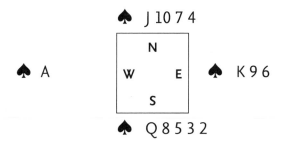

♠ J 10 7 4

♠ A   N W E S   ♠ K 9 6

♠ Q 8 5 3 2

Covering would be disastrous on this third layout too.

Devalue 4333 hands

Hands with three three-card suits and one four-card suit are bad news. Bridge players use various adjectives to describe such hands – 'flat' and 'sterile' for example.

These hands offer little trick-taking potential. In a no-trump contract, only one suit offers the chance of developing long cards. In a suit contract, you cannot normally ruff anything.

4333 hands are also bad defensively because they mean that suits in which the opponents hold seven or eight cards are breaking kindly for them.

If ever you have a marginal bidding decision, you should take the conservative view if you have a 4333 hand.

If you show 12-14 points (by opening 1NT if you are playing a weak no-trump) and partner invites you to bid game by raising to 2NT, you should decline.

Don't double into game

Contracts below 3NT (as well as 4♣ and 4♦) are not normally game contracts unless someone doubles. If, however, you double the opponents in any contract of 2♥ or higher, they will make game if the contract succeeds. You have a lot to lose.

If you turn 50 into 100 by doubling, you have gained next to nothing. By contrast, if you turn a loss of 110 into a loss of 470 (duplicate scoring), you have lost a lot. In that scenario, you are risking 360 to gain 50. You would have to be right almost nine times out of ten to show a profit.

The odds are slightly better when the opponents are vulnerable. At match-point scoring, the difference between 100 and 200 could be significant if most people are making a part-score your way. Even then, it can be hard to judge that they are going exactly one down – if they are going two down, there may be no need to double to get a good score. The bottom line is this: if you are in any doubt at all, do not double if in doing so you change their part-score into a game contract.

Don't bid your hand twice

Once you have described your hand, you should usually shut up and let partner make the decisions. While it is true that your partner or the opponents may be in a contract that you do not fancy, this is no excuse for showing the same features twice.

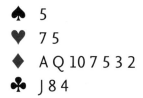

♠ 5
♥ 7 5
♦ A Q 10 7 5 3 2
♣ J 8 4

If you open 3♦ and partner responds 3NT, you pass. You do not worry about the singleton spade or that you have a long suit. Your 3♦ already showed what you have.

♠ A J 9 5 2
♥ Q
♦ A Q 10 7
♣ 7 4 2

Suppose you open 1♠, partner responds 1NT, you rebid 2♦ and your partner rebids 2♥; you pass. You have already shown at least five spades and four diamonds, so partner will not be expecting you to have many hearts.

If you make any sort of limit bid (such as a 1NT opening or overcall), again the onus is on you to let partner take the decisions. You have already shown your points.

Risky ruff and discard

As a defender, it is usually a mistake to lead a suit when you know that both declarer and dummy are void. If you lead the suit, declarer can ruff in one hand (normally the short trump hand) and throw a loser from the other.

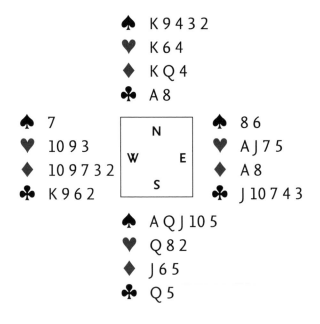

♠ K 9 4 3 2
♥ K 6 4
♦ K Q 4
♣ A 8

♠ 7
♥ 10 9 3
♦ 10 9 7 3 2
♣ K 9 6 2

♠ 8 6
♥ A J 7 5
♦ A 8
♣ J 10 7 4 3

♠ A Q J 10 5
♥ Q 8 2
♦ J 6 5
♣ Q 5

South plays in 4♠. You, West, lead the ♦10. Partner takes the ♦K with the ♦A and switches to the ♣J. The ♣Q and ♣K cover this, the ace winning. Declarer takes two trumps and two diamonds before exiting with a club to your nine.

If you lead a diamond, you give a ruff and discard. While the club position is less clear, the opponents could both be void. You do not want declarer to ruff in one hand and throw a heart from the other. Switch to the ♥10.

Not overruffing

As a defender, making declarer ruff in front of you can be a good tactic. If you get the chance to overruff cheaply, it is usually right to do so. If, however, you would have to overruff with a winner, you may do better to discard. This can promote a lower card to winning rank.

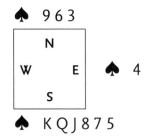

♠ 9 6 3

N
W   E
S

♠ A 10 2   ♠ 4

♠ K Q J 8 7 5

This is the trump suit. East leads some suit in which both West and South are void. If South ruffs with an honour and West overruffs with the ace, West's remaining trumps fall under South's other two honours. Instead, West should discard, later making both the ace and ten.

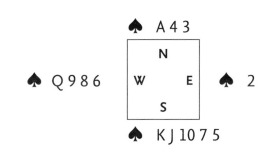

♠ A 4 3

♠ Q 9 8 6

♠ 2

♠ K J 10 7 5

This is similar. West should not overruff the jack or ten with the queen. Better is to save the queen and make two trump tricks later.

Don't underlead an ace

At trick one in a suit contract, you should not underlead an ace (nor should you lead an ace – the role of an ace is to capture honours, not low cards). One risk in underleading an ace is that declarer may have a king in one hand and a singleton in the other, in which case the ace never makes.

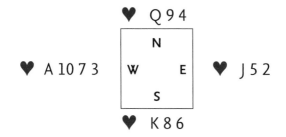

♥ Q 9 4

N
W   E
S

♥ A 10 7 3   ♥ J 5 2

♥ K 8 6

Leading from a suit with the ace can also blow a trick if declarer has honours in each hand. If you lead this suit, the king and queen both make for declarer.

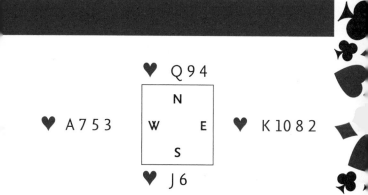

♥ Q 9 4

♥ A 7 5 3   N W E S   ♥ K 10 8 2

♥ J 6

Underleading an ace may also confuse partner. If you lead the three and dummy plays low, East will naturally finesse the eight, thinking you have the jack and declarer the ace rather than the other way around. Then declarer loses only one heart trick instead of two.

No-trump penalty double

If an opponent opens 1NT, a double is for penalties rather than for takeout. When the opponents have not bid any suits, you cannot really have support for all the other suits! How strong you need to be depends upon the strength of the 1NT opening. If the 1NT shows 12-14 (a weak no-trump), you need at least 15 points. If 1NT shows 15-17 (strong), you need at least 17 points. You might shade these by a point if you have a good suit to lead. Once you double 1NT, if the opponents run to a suit, subsequent doubles are for penalties too. You do not want to let them off the hook!

Hand 1
♠ A Q 5
♥ Q J 10 8
♦ K J 9
♣ Q 8 4

Hand 2
♠ K Q 10 6 5
♥ 8
♦ K J 9 3
♣ A K Q

Hand 1 is good enough to double a weak no-trump. You intend to lead the ♥Q. With Hand 2, you can double a weak or a strong no-trump. You do not promise a balanced hand when you double 1NT. You intend to compete with 2♠ if, as is quite likely, the opponents run to 2♥.

Note an important exception to the rule that no-trump doubles are for penalties. If one opponent opens a suit and the other responds 1NT, double is a takeout double of opener's suit.

Eight ever, nine never

If you are missing the queen in a suit and need to judge whether to finesse or play for the drop, you finesse if you have eight cards between the two hands but not with nine.

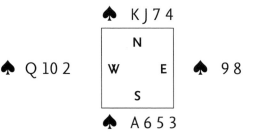

You should cash the ace and finesse the jack. Since the queen is more likely to be in a tripleton than a doubleton, the finesse is clearly the better play.

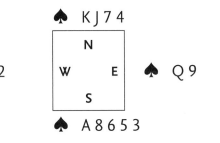

♠ K J 7 4

♠ 10 2 ♠ Q 9

♠ A 8 6 5 3

If you add a ninth card, the odds are quite close, though you will do better if you never finesse than if you always do. Here, if you cash the ace and play a second round, by the time West follows twice, you will know about two cards in West's hand and one in East's. This leaves a bit more space in the East hand for the missing queen. The original chance of finding West with Q-x-x is in fact 18%, which compares to a 20% chance of finding East with Q-x.

Four-deal bridge

In traditional rubber bridge, it can take a long time for one side to make the two games required to complete a rubber. This can be frustrating if someone has to leave at a fixed time or if you have a player sitting out.

A popular form of the game nowadays is four-deal bridge (also known as Chicago). You play exactly four deals. Each player deals once. On the first deal, neither side is vulnerable. On the fourth deal, both sides are vulnerable. On the middle two deals, one side only is vulnerable (most play it as the dealing side).

In four-deal bridge, you get a bonus for game when you bid it, 300 non-vulnerable and 500 vulnerable. You can carry forward part-scores as at rubber bridge and there is a 100 bonus for a part-score made on the fourth deal.

Penalties are the same as at other forms of the game. Non-vulnerable undoubled undertricks are 50 a time; vulnerable undoubled undertricks are 100 a time.

When doubled and non-vulnerable, undertricks are 100 for the first undertrick, 200 for the second and third, 300 for any more undertricks. When doubled and vulnerable, undertricks are 200 for the first and 300 for each subsequent undertrick.

Honours count as in rubber bridge, 150 for the AKQJ10 of trumps or four aces in no-trumps, 100 for four trump honours.

Defending actively

As a defender, one of the most difficult tasks to judge is whether to sit back and hope declarer runs out of steam or whether to open up new suits in the hope of setting up or cashing winners. If dummy has an obvious source of tricks, such as a long strong suit, you should get active, attacking new suits where declarer might be weak.

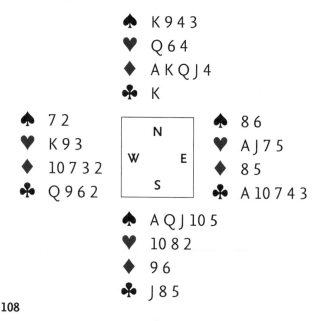

South plays in 4♠. West leads the ♣2, which you win with the ♣A. The diamond position seems ominous. You do not want declarer to get in, draw trumps and run the diamond suit.

You should get active, switching to the ♥5. You hope West has the king but no more than three hearts in total.

Hold up a stopper

If it makes no difference to how many tricks you make in a suit, you usually hold up a stopper. This applies mainly to declarer, though the defenders can employ the tactic too.

```
              ♠  A 9 4
              ♥  10 6
              ♦  A J 7 2
              ♣  K Q 5 4

♠  7 2              N            ♠  K 8 6 3
♥  K J 9 5 3                     ♥  Q 7 4
♦  K 8 3     W          E        ♦  Q 9 5
♣  9 6 2           S            ♣  10 7 3

              ♠  Q J 10 5
              ♥  A 8 2
              ♦  10 6 4
              ♣  A J 8
```

As South you play in 3NT. West leads the ♥5. If you take East's queen with the ace at trick one, disaster strikes after the spade finesse loses. East returns a heart, causing you to lose four heart tricks and a spade.

Now see what happens if you hold up the ♥A, waiting to play it on the third round. This time East has no hearts left to play after getting in with the ♠K.

Holding up would also be right with A-x-x facing 10-x-x.

Declarer, with more strength between the two hands, is less likely to have communication difficulties, which is why a hold up is less often successful for the defending side. Nevertheless, it can be useful technique, especially when most of declarer's strength is in one hand.

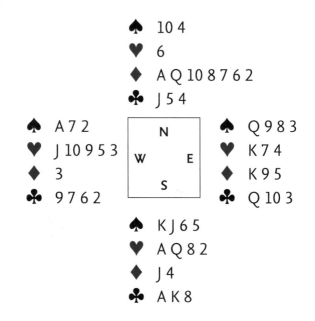

♠ 10 4
♥ 6
♦ A Q 10 8 7 6 2
♣ J 5 4

♠ A 7 2
♥ J 10 9 5 3
♦ 3
♣ 9 7 6 2

N
W E
S

♠ Q 9 8 3
♥ K 7 4
♦ K 9 5
♣ Q 10 3

♠ K J 6 5
♥ A Q 8 2
♦ J 4
♣ A K 8

South plays in 3NT. West leads the ♥J. Your king loses to the ace. Declarer leads the ♦J and West plays the ♦3.

If you take the king on the first round, declarer romps home, making six diamonds, two hearts and two clubs for an overtrick. It is a different story if you hold up the ♦K for one round. This shuts out the long diamonds, restricting your opponent to just two tricks from the suit.

When defending, you should aim to keep a card to lead in partner's suit, especially if partner has ready winners in it.

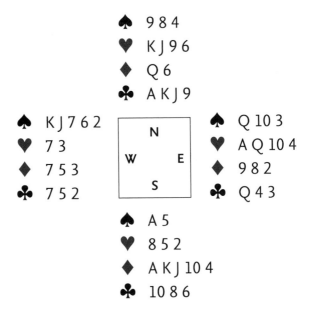

♠ 9 8 4
♥ K J 9 6
♦ Q 6
♣ A K J 9

♠ K J 7 6 2
♥ 7 3
♦ 7 5 3
♣ 7 5 2

♠ Q 10 3
♥ A Q 10 4
♦ 9 8 2
♣ Q 4 3

♠ A 5
♥ 8 5 2
♦ A K J 10 4
♣ 10 8 6

South plays in 3NT. West leads the ♠6. You, East, put up the queen, which wins. Declarer wins the second spade and runs five rounds of diamonds, throwing a spade and two hearts from dummy. You must find two discards.

While declarer may be planning to take the club finesse, you want to keep the queen guarded if you can. You also need to keep a spade to play to partner's winners or declarer can get home with a spade trick, five diamonds and three clubs. You should discard two hearts.

Keep enough winners

As a defender, when making a discard, you often have to make a choice. Do you keep your own winners, or should you retain stoppers in declarer's suits? If you can, keep enough winners to defeat the contract. This is especially so in no-trump contracts – or if declarer is short of trumps.

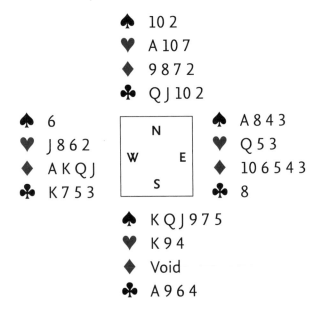

♠ 10 2
♥ A 10 7
♦ 9 8 7 2
♣ Q J 10 2

♠ 6
♥ J 8 6 2
♦ A K Q J
♣ K 7 5 3

♠ A 8 4 3
♥ Q 5 3
♦ 10 6 5 4 3
♣ 8

♠ K Q J 9 7 5
♥ K 9 4
♦ Void
♣ A 9 6 4

South plays in 4♠. You lead a top diamond, ruffed. A trump goes to the ten and ace. Partner makes declarer ruff a second diamond. Declarer draws three more rounds of trumps, forcing you to make three discards. Two are easy – two low clubs. The third had better not be a diamond. You need to keep two diamonds to make when you come in with the ♣K to give you four tricks. So throw a heart.

Keep length with dummy

When defending, it can be hard to find safe discards. One suit worth protecting is a suit in which dummy has length (usually four cards). Here East must keep four diamonds:

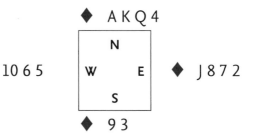

◆ A K Q 4

◆ 10 6 5   ◆ J 8 7 2

◆ 9 3

The next layout is essentially the same, though it is now less obvious that East has a potential fourth-round winner:

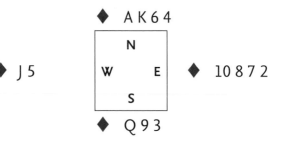

◆ A K 6 4

◆ J 5   ◆ 10 8 7 2

◆ Q 9 3

If the layout below is a side suit in a suit contract, one of the defenders needs to keep four diamonds here as well:

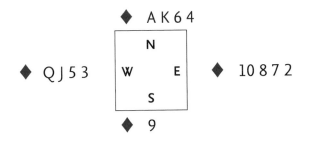

One defender must keep four diamonds or declarer can set up a long diamond by a ruff.

Know the odds

Although bridge is not a very mathematical game – most of the time all you need is the ability to count to 13 – it does help to know a few simple odds.

Distribution of 2 missing cards
1-1	52%
2-0	48%

Distribution of 3 missing cards
2-1	78%
3-0	22%

4 missing cards		5 missing cards	
2-2	40%	3-2	68%
3-1	50%	4-1	28%
4-0	10%	5-0	4%

6 missing cards			
3-3	36%	One finesse	50%
4-2	48%	One of two finesses	75%
5-1	14%	Two of two finesses	25%
6-0	2%	Three finesses	13%
		Two of three finesses	50%

Remember, these are the odds at the start of the deal when you have no information about the unseen hands.

It is worth remembering that an odd number missing will usually break as evenly as possible whereas an even number will usually not divide exactly in half.

Lead low

As declarer, you often start a suit by leading low, saving high cards for when the opposing high cards have gone.

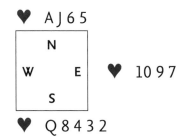

♥ A J 6 5

♥ K

♥ 10 9 7

♥ Q 8 4 3 2

If you lead the queen, you need a 2-2 split with the king right. Leading low gives the extra chance of a bare king.

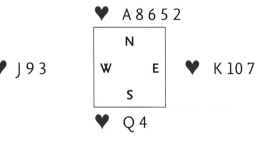

♥ A 8 6 5 2

♥ J 9 3

♥ K 10 7

♥ Q 4

If you lead the queen, it never makes. Instead, you make the first lead from the North hand, low towards the queen.

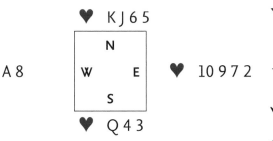

♥ K J 6 5

N
W   E
S

♥ A 8   ♥ 10 9 7 2

♥ Q 4 3

If you lead the queen, you will need a 3-3 break to make three tricks. Instead lead low (twice) towards dummy, obliging West to play the ace on a low card.

Give nothing away

Suits with honours scattered around the table are often dangerous to play. Whichever side plays the suit first often blows a trick. While it is obvious not to lead into a simple tenace, leading into a split tenace can be just as costly.

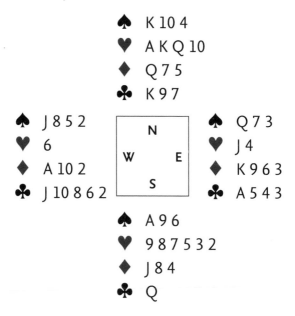

South plays in 4♥. West leads the ♣J. East wins with the ace, dropping South's queen.

A club return into the jaws of the K-9 would clearly be a bad idea. The danger signs are also there in the pointed suits (spades and diamonds). East has honours in those suits and can see the honours in dummy. The only safe return is a trump. This way declarer has to make the first play in the frozen suits and ends up a trick short. Leading any of the side suits would give away the contract.

Don't draw winning trumps

When you are drawing trumps and get to the point where the defensive trumps are winners, you usually do best not to play any more trumps. You do not want to play two of your trumps, one from hand and one from dummy, to take out one opposing trump. Another reason to leave the master trump out is that if you give up the lead you give a tempo, allowing the defenders to cash or set up a winner.

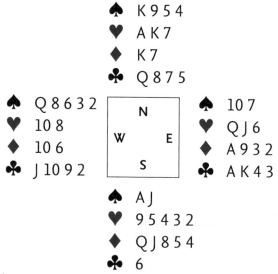

♠ K 9 5 4
♥ A K 7
♦ K 7
♣ Q 8 7 5

♠ Q 8 6 3 2
♥ 10 8
♦ 10 6
♣ J 10 9 2

♠ 10 7
♥ Q J 6
♦ A 9 3 2
♣ A K 4 3

♠ A J
♥ 9 5 4 3 2
♦ Q J 8 5 4
♣ 6

South plays in 4♥. West leads the ♣J followed by the ♣10. Declarer ruffs, knocks out the ♦A and ruffs the next club. It is right to cash two top trumps (to stop West from ruffing a diamond) but then you leave the master trump out. You want to keep a trump in dummy for ruffing a diamond and one in hand for dealing with any more clubs.

Make the long hand ruff

If you are defending, sometimes a safe way to get off play is to play a suit in which dummy is weak and declarer is void: this gives nothing away. A more dynamic reason for making the long hand ruff is to reduce declarer to holding fewer trumps than you hold yourself.

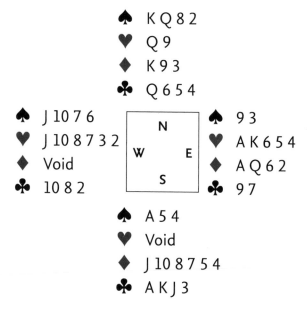

♠ K Q 8 2
♥ Q 9
♦ K 9 3
♣ Q 6 5 4

♠ J 10 7 6
♥ J 10 8 7 3 2
♦ Void
♣ 10 8 2

N
W　E
S

♠ 9 3
♥ A K 6 5 4
♦ A Q 6 2
♣ 9 7

♠ A 5 4
♥ Void
♦ J 10 8 7 5 4
♣ A K J 3

South plays in 5♦. West leads the ♥J. Declarer ruffs and runs the ♦J. East wins with the queen and continues hearts, making the long hand ruff a second time. Declarer then plays a second round of diamonds. If East wins this, a third round of hearts will not make the long hand ruff – dummy would ruff instead. East therefore holds up the ace until the third round of diamonds, poised to make South ruff again with a third round of hearts.

Nine is easier than eleven

Nine tricks are usually easier to make than eleven, even when a trump suit might give you an extra winner in a suit contract. When you are deciding whether to bid game in a minor or to stop in 3NT, this is useful guidance.

♠ 8 4
♥ K 6 3
♦ J 7
♣ A K J 8 5 2

If partner shows 12-14 balanced, you should simply raise to 3NT. Do not even think about playing in clubs.

♠ A 5
♥ A 6
♦ A K 10 7 4 2
♣ Q 8 3

If you open 1♦ and partner makes a simple raise (or a pre-emptive double raise), you should simply bid 3NT. You expect the diamonds to run, which gives you eight certain tricks. Anything of value in partner's hand should give you a ninth (the ♠K or ♥K for example). By contrast, 5♦ is a long way off. Partner will need several useful cards for you to come to eleven tricks.

At match-point scoring an added reason for preferring 3NT is that 3NT with an overtrick outscores 5♣ or 5♦ just making.

Obey the rules

Sometimes players do things that are not in accordance with the correct procedure. They bid when it is not their turn, they lead when it is not their turn – or they fail to follow suit when they could. Players are only human and do occasionally do such things! Rules exist to deal with these sorts of situation and others. If you are playing in a club, you should ask for the director or host to come to your table, who will then give the appropriate ruling.

If you are playing at home, you should consult a rulebook if you can find one. It explains the penalties and options available. These can be quite complicated. The rules on a bid out of turn depend upon whose turn it really was to bid. The penalties on a revoke (failure to follow suit) depend upon what tricks the offending side won subsequently and on how quickly you noticed the revoke. The remedies are also a bit different depending upon whether declarer or one of the defenders committed the infraction.

Very few people know all of the rules and regulations. This is why it is important to call for the director if you can.

Play low in the second seat

This is a rule for both sides. Let us start with declarer:

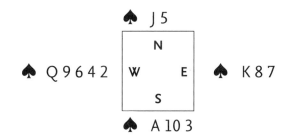

```
            ♠ J 5
           ┌─────────┐
           │    N    │
♠ Q 9 6 4 2│ W     E │♠ K 8 7
           │    S    │
           └─────────┘
            ♠ A 10 3
```

If West leads the four, you ensure two spade tricks (a double stopper) if you play low from dummy.

```
            ♠ K 5
           ┌─────────┐
           │    N    │
♠ Q 9 6 4 2│ W     E │♠ 10 8 7
           │    S    │
           └─────────┘
            ♠ A J 3
```

Again West leads the four. This time you ensure three spade tricks if you play low from dummy.

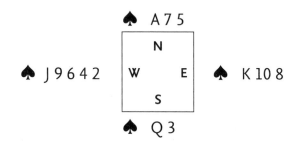

♠ A 7 5

♠ J 9 6 4 2   N W E S   ♠ K 10 8

♠ Q 3

Again West leads the four. Playing dummy's ace means the queen never makes. By contrast, if you let the lead run to your hand, you ensure two spade tricks. Playing low in second seat is usually right if you have strength in the fourth seat and the high cards in the second seat might win later.

As a defender, you cannot be sure what partner has. All the same, playing second hand low is usually right.

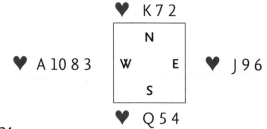

♥ K 7 2

♥ A 10 8 3   N W E S   ♥ J 9 6

♥ Q 5 4

If South leads low and West flies in with the ace, the king and the queen will both score. West should play second hand low to save the ace for capturing the queen.

♥ J 7 2

♥ K 10 8 3

♥ A 9 6

♥ Q 5 4

If South leads low and West goes in with the king, the queen and jack become equals for driving out East's ace.

♥ A 9 2

♥ J 8 3

♥ Q 7 6 5

♥ K 10 4

If South leads low and West plays the jack 'to force out the ace', declarer can finesse the ten on the way back. There is no need to play high on such layouts. Partner's presumed high cards will stop declarer f rom making a cheap trick.

Pre-empt to silence them

When you have a weak hand and either a very long suit of your own or a good fit for your partner's suit, you should bid to the limit, or 'pre-empt', straight away. Even if you go down, you will probably stop the opponents from getting a good score.

♠ A Q J 8 7 5 2
♥ 4
♦ 8 4
♣ 7 5 2

You should open 3♠.

♠ 5
♥ 4
♦ 8 4 2
♣ A K J 9 8 7 5 2

You should open 4♣.

♠ 5
♥ Q 10 7 4 2
♦ 8 4
♣ K J 9 8 7

If partner overcalls in hearts, you should jump to 4♥.

Pre-empting forces the opponents to bid (if they want to) at a high level. Sometimes they will guess wrong.

Lead majors not minors

As opening leader, if it is unclear what suit to lead, you should generally prefer an unbid major to an unbid minor. Why is this? If the opponents have the majors, they will tend to bid them in the hope of finding an eight-card fit.

West	North	East	South
			1NT
Pass	3NT	End	

North surely does not have a five-card major (no transfer or jump to three of a suit) and probably no four-card major either (no Stayman).

West	North	East	South
			1♦
Pass	3♦	Pass	3NT
End			

North does not have a four-card or five-card major but could have length in clubs.

West	North	East	South
	1♦	Pass	1NT
Pass	3NT	End	

South does not have a four-card major and probably does not have four diamonds either. A speculative club lead is likely to hit South's main suit on this auction.

Ruff high if you can

When you know or suspect that an opponent can overruff, ruff as high as you can afford. You may need to save your very highest cards to draw the opposing trumps, in which case you ruff with a high spot card. If, however, you have a surplus of high cards for drawing them, you ruff very high.

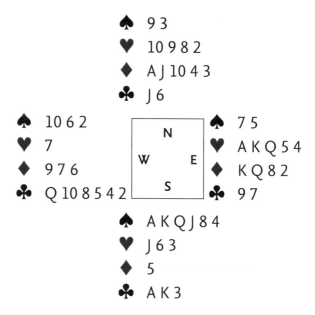

```
                    ♠  9 3
                    ♥  10 9 8 2
                    ♦  A J 10 4 3
                    ♣  J 6
    ♠  10 6 2                        ♠  7 5
    ♥  7               N             ♥  A K Q 5 4
    ♦  9 7 6       W       E         ♦  K Q 8 2
    ♣  Q 10 8 5 4 2       S          ♣  9 7
                    ♠  A K Q J 8 4
                    ♥  J 6 3
                    ♦  5
                    ♣  A K 3
```

South plays in 4♠. The defenders start with four rounds of hearts. Declarer should ruff with an honour because it is much more likely that West holds 10-x-x or 10-x of trumps than that East has 10-x-x-x. Moreover, when declarer goes on to ruff the third round of clubs in dummy, this should be with the nine; even though the clubs are 6-2, there is then the chance that West holds the ten of trumps.

The rule of eleven

When you lead a low card, have you ever wondered why you should lead precisely the fourth highest? You do so to give your partner information. If you have three cards higher than the one you have led, partner can look at dummy and work out how many higher cards declarer has. Rather than do this from first principles, you use the rule of 11. You deduct the size of the spot card led from 11. This says how many higher cards there are in the other hands.

```
                  ♠  K 10 7
              ┌─────────────┐
              │      N      │
♠  J 9 6 5 2  │  W       E  │  ♠  Q 8 4
              │      S      │
              └─────────────┘
                  ♠  A 3
```

West leads the five. East takes five from eleven to give six. Five of those six are visible to him – the king, ten and seven in dummy as well as the queen and eight. East thus places South with only one higher than the five. In a suit contract, this must be the ace (as West would not have underled the ace).

The size of the spot card also reveals length.
If the lead is the two, it is from only a four-card suit.
If the lead is the three and the two is visible, again it is from only a four-card suit; if the two is missing, the lead might be from a five-card suit.

Note that declarer (to work out what the leader's partner has) can use the rule of eleven too!

Suit-preference signal

When partner does not need to know your length or strength in a suit, it is usual to give a suit-preference signal. The most common situation for this is when giving partner a ruff. You know that the same suit is not coming back. The way you give a suit-preference signal is that you lead a high card to ask for the higher-ranking suit and a low card for the lower-ranking suit. You can exclude the trump suit from the equation.

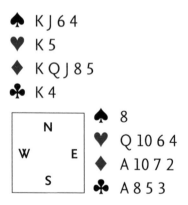

♠ K J 6 4
♥ K 5
♦ K Q J 8 5
♣ K 4

♠ 8
♥ Q 10 6 4
♦ A 10 7 2
♣ A 8 5 3

South plays in 4♠. West leads the ♦3, clearly a singleton.

Which diamond should you return after taking the ♦A? Since your entry is in the lower-ranking club suit, you return the ♦2. Partner then ruffs the diamond, puts you back in with the ♣A and scores a second ruff when you play a third round of diamonds.

If you had the ♥A instead of the ♣A, you would return the ♦10, your highest remaining diamond to ask for the higher suit.

Stay low on a misfit

On many hands, the trick-taking potential depends upon having a trump fit. When you have no eight-card fit anywhere, you often cannot make very much. If you have K-Q-J-x-x in one suit and K-Q-x-x-x in another, this looks good. However, if partner has similar holdings in the other two suits, you may not take many tricks. Indeed, on a misfit deal, you may find that you make even fewer tricks than you would if you had two balanced hands – you will have difficulty crossing from one hand to the other.

♠ 9 4 3
♥ 2
♦ K 6 5
♣ K J 8 5 4 2

Partner opens 1♥ and rebids 2♦ over your 1NT response. You should stay low on the misfit and pass.

♠   Void

♥   A K 8 2

♦   K 6 5 3

♣   K J 8 5 4

Partner opens a weak 2♠, showing six spades and 5-9 points What should you do? Even though you expect to have only six spades between you, the right action is to pass. Any action would take the bidding higher with no assurance of finding a fit.

The card to switch to

If, as a defender, you lead a new suit during the play, the lead follows the same principles as an opening lead: top of a sequence, fourth highest from a long suit and so on.

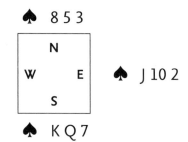

♠ 8 5 3

♠ A 9 6 4

♠ J 10 2

♠ K Q 7

East switches to the jack of spades. If South plays the king, West can place East with the ten and South with the queen.

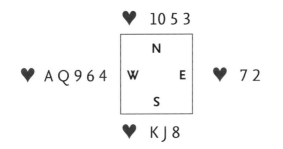

♥ 10 5 3

♥ A Q 9 6 4

♥ 7 2

♥ K J 8

East, if switching to a heart, should lead the seven. West can then make two heart tricks and, seeing East follow with a lower card, place declarer with the third heart. If this is a side suit, West can then give East a ruff.

Note that on some hands common sense tells you what to do. For example, if dummy has a singleton queen and you hold K-J-10, you switch to the king, pinning the queen.

Transfers over 1NT

If partner opens 1NT, a very useful convention is transfers. If you play transfers, a bid of 2♦ shows hearts (the suit above the suit you bid) and a bid of 2♥ shows spades. The main reason for playing transfers is that you get a second chance to bid, enabling you to use a transfer on both weak and strong hands. The other benefit is that the 1NT opener, who is more likely to hold tenaces, becomes declarer if you play in your major.

Hand 1
♠ 8 5
♥ Q J 10 9 3
♦ A 8 4 3
♣ 7 3

Hand 2
♠ 8 5
♥ A J 8 4 3
♦ K Q 6
♣ J 10 3

On either hand, if partner opens 1NT, you start with 2♦, which asks partner to bid 2♥.

With Hand 1, you do not wish to go any higher than 2♥, though it could be beneficial to have the lead coming up to partner's possible tenaces, such as the ♠A-Q or the ♣K.

With Hand 2, (assuming 1NT shows 12-14 points) you want to invite game while showing five hearts and a balanced hand. With transfers, you can do it. You bid 2♦ and then rebid 2NT. With a minimum, partner then passes or bids 3♥; with a maximum, partner can raise to 3NT or jump to 4♥.

Two opening bids produce game

If you have enough for a one-level opening and partner does too, the odds are that you have enough between you to make game. This is obvious when the two hands are balanced. You will have an absolute minimum of 12 points facing 12; most of the time you will have more than that.

If you know that you want to reach game, either you bid game if you know which game you want to play in or make a forcing bid if you do not.

♠	Q 6
♥	A K 9 6 4
♦	K J 10 3
♣	7 2

Suppose partner opens 1♠ and rebids 2♠ over your 2♥ response. You do not want to commit to what might be a 5-2 spade fit. Nor do you want to bid 3NT with the clubs possibly wide open. However, you do want to reach game as you have an opening bid facing an opening bid. Thus, you bid 3♦.

Now suppose that partner opens 1♥. In this case, you know you want to play in game (at least) and that you want to play in hearts. You should bid 2NT if you play that as a game-forcing heart raise. If you do not have a way to show a strong heart raise, you should start with 2♦ and support hearts vigorously later.

The losing trick count

On many deals, valuing your hand in terms of (Milton) points with appropriate adjustments for long or short suits is all you need. When you have a fit, however, you may find it quicker and as accurate to use the losing trick count.

You count at most three losers in a suit and deduct losers for aces, kings and normally queens.

No losers: Void, A alone, A-K alone, A-K-Q to any length

One loser suits: Singleton K or lower, K-x or A-x, A-K-x, A-Q-x, K-Q-x

Two loser suits: Q-x or worse, A-x-x, K-x-x, Q-J-x

You count Q-x-x as three losers unless you have an ace elsewhere or partner has bid the suit.

Hand 1
- ♠ K 9 6 4
- ♥ A 5
- ♦ K Q 10 2
- ♣ 8 6 4

Hand 2
- ♠ J 9 6 4
- ♥ Q
- ♦ A Q 9 5 2
- ♣ 8 6 4

Hand 1 contains seven losers: two in spades, one in hearts, one in diamonds and three in clubs.

Hand 2 contains eight losers: three in spades, one in hearts, one in diamonds and three in clubs.

Typical loser counts for various bids are as follows:

Game-forcing opening = 3 or fewer losers

Strong but not game-forcing opening = 4 losers

Strong 1-level opening or jump shift response = 5 losers

Good 1-level opening or minimum jump shift = 6 losers

Minimum opening or game values as responder = 7 losers

2-level response or limit raise of an opening = 8 losers

1-level response or single raise of an opening = 9 losers

Now comes the clever bit. To work out how many tricks the partnership can make, you add your losers to partner's losers and deduct the answer from 24. So, if you have a spade fit and both have seven-loser hands – you can probably make game in spades (24–7–7=10).

On Hand 1, with its seven losers, if partner opens 1♠, you would expect to make 4♠. You will make your system bid to show a game-forcing raise if you have one or make some forcing bid before bidding 4♠ next time.

On Hand 2, with its eight losers, you would not be so sure of making game if partner opens 1♠. You would just make a limit raise to 3♠, inviting partner to go on to game with a six-loser hand or to pass with seven losers.

Note well that the losing trick count only works when you have a fit and can exaggerate the potential on some deals – so do not bid a slam every time you have two six-loser hands facing one another!

Lead second from bad suits

When you have a sequence of honours, you lead an honour. When you have an honour or two but not a sequence, you lead fourth highest. What do you lead from a bad suit, a suit you probably do not want partner to return? The standard lead is the second highest card. From 8-6-3, you lead the six; from 9-7-5-2, you lead the seven.

How does partner tell that the lead is second highest? One way is to try using the rule of 11. If the rule of 11 fails – because there are too many higher cards about, partner can work out that the lead is second highest.

From a three-card suit, it is customary to play the middle card first, then the upper (higher) card and finally the lowest card (down). This strategy of leads goes by the mnemonic Middle-Up-Down or MUD. You play the uppermost card second to differentiate the holding from a doubleton.

Note that in the days of whist the norm was to lead the highest card from rubbish, the 'top of nothing'. Some players still prefer this method. If you have raised the suit – so that partner knows you do not have a doubleton anyway – it is also quite common for a partnership to agree to lead the top rather than the middle card from a collection of low cards.

Valuable intermediates

The Milton point count method does not explicitly allow for intermediate cards – high spot cards such as tens and nines. You therefore need to make an adjustment on hands particularly weak or strong in intermediates. The average hand has one ten and one nine. If you have no tens or nines, you should take off a point. If you have two tens and two nines, you should add a point.

Why are intermediates so valuable? You need to consider their impact in the play. Suppose that you have A-Q-2 facing 5-4-3. You will make two tricks if the finesse works, one if it does not. On average, you make one and a half tricks. Now suppose you have A-Q-10 facing 5-4-3. You will make three tricks if the king and jack are both onside, two tricks if only one of those cards is right. On average, you make two tricks, a half trick better.

Hand 1
- ♠ A J 10 5
- ♥ Q J 10 8
- ♦ A Q 9
- ♣ K 9

Hand 2
- ♠ A J 4 2
- ♥ Q J 6 3
- ♦ A K 2
- ♣ K 2

You should add a point for the good spot cards on Hand 1 but deduct a point for the poor spot cards with Hand 2. Hand 1 is too good for a 15-17 1NT rebid (or opening). Hand 2 is not too strong for an action showing 15-17 points.

Watch your entries

Watching your entries is crucial. It is all very well setting up winners – reaching them is another matter. Take extra care when you start with one or more blocked suits:

♠ A 7 5 4
♥ K 4
♦ K Q 3
♣ Q 7 5 3

♠ K Q J
♥ A 8 3
♦ J 10 7 6
♣ 10 6 2

As West, you play in 3NT. North leads the ♥Q. Suppose you win with the ♥A. If you then cash the ♠K-Q-J, you will make only two diamond tricks if the opponents hold up the ace until the third round. Instead, you should win with the ♥K, unblock the spades and lead a diamond. The ♦K-Q will provide a sure entry for cashing the ♠A, while the ♥A gives an entry to dummy's long diamond.

♠ 9 8 7 5 4
♥ A 9 4 2
♦ A 6
♣ 7 5

♠ K Q J 10
♥ 8 3
♦ J 8 7 3
♣ A Q 8

You, West, are in 3♠. North leads the ♥Q.
If you win and concede a heart, the defenders
may play ace and another trump. Although you
can ruff a heart, come to the ♦A and ruff a heart,
you will lack a quick entry to ruff another heart, and
so may lose the ruff. So duck the first trick, making
the ♥A an entry.

Leading an unbid suit

The opponents bid suits in which they have length and strength. This means you should lead suits that they have not bid. As the trump suit is always a bid suit, only lead a trump when the bidding calls for a trump lead. Remember, as declarer, you often play to draw trumps. You would not expect it to be right for both sides to lead the same suit.

West	North	East	South
			1♠
Pass	2♦	Pass	2♠
Pass	3♠	Pass	4♠
End			

♠ 10 5
♥ A J 6
♦ Q J 9 3
♣ Q 10 6 3

The opponents have bid the pointed suits, while hearts with the unsupported ace is an unattractive lead. Lead the ♣3.

♠ 6 4
♥ K 9 6 3
♦ A K 5
♣ J 10 8 2

You do not want to set up the diamonds or sit back waiting for declarer to do so. Lead the ♣J.

Aces facing singletons

When it comes to a high-level bidding decision – whether to play or defend – or whether to go for a slam or not – how well your side's hands fit can be crucial.

If partner has a singleton somewhere, the best holdings for you are A-x-x or a collection of low cards. The king or the queen may be of no use offensively.

How do you know when partner holds a singleton? If the opponents bid a suit vigorously and you hold three cards in the suit, partner is likely to have a singleton. A partner who bids three suits should have a singleton in the fourth. You can also tell when partner makes a conventional bid to show a singleton, such as a splinter bid.

Hand 1

♠ A Q 9 8 3
♥ K 7 4
♦ K Q 6
♣ Q 3

Hand 2

♠ A Q 9 8 3 2
♥ K 4
♦ A 8 6 4
♣ 5

Suppose you open 1♠ and partner responds 4♦, showing a raise to 4♠ with a singleton diamond. With Hand 1, you know the hands fit badly, with the king-queen facing the singleton. With Hand 2, by contrast, you know the hands fit very well. A sign off in 4♠ is right with Hand 1. With Hand 2 you should check on key cards with a view to reaching 6♠ – there should be no diamond losers.

Add three points when protecting

When one opponent opened the bidding and the other opponent passes, it is reasonable to assume that you and your partner have fair values between you – otherwise they would not be stopping at the one level. In the fourth seat, you can afford to bid a little lighter than usual because you know that your partner must have something. How much lighter can you be? The customary rule is that you can bid with about three points (a king) less than you would need in the second seat.

West	North	East	South
	1♠	Pass	Pass
?			

Hand 1
- ♠ 6
- ♥ Q 10 4 2
- ♦ J 9 5 4
- ♣ A Q 6 3

Hand 2
- ♠ A 10 6
- ♥ K 7 4
- ♦ Q J 6 4
- ♣ K 10 3

With Hand 1, if you add 3 points to your actual 9, you have 12 – enough for a reopening double.

With Hand 2, if you add 3 points to your actual 13, you have 16 – enough for a 1NT overcall.

Of course, if you are the one who has passed in the second seat, you need to bear in mind that your partner has already bid 3 of your points, tailoring your actions accordingly.

Fear frozen suits

A frozen suit has honours spread around the table and costs a trick to whoever attacks it.

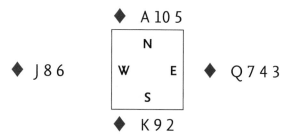

♦ A 10 5

♦ J 8 6   ♦ Q 7 4 3

♦ K 9 2

Neither defender can afford to attack this suit. If you do have to lead it, best is to lead an honour, pretending to hold the queen and jack.

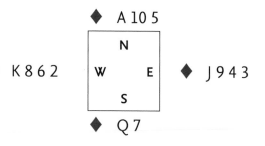

♦ A 10 5

♦ K 8 6 2   ♦ J 9 4 3

♦ Q 7

Again, neither defender wants to lead the suit. It is slightly better if East does because declarer might guess wrong (putting up the queen, playing East for the king).

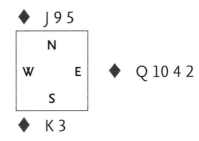

♦ J 9 5

♦ A 8 7 6   ♦ Q 10 4 2

♦ K 3

Neither defender wants to lead the suit – but if someone has to it should be East, giving declarer a guess.

Bid to your trump number

Trumps take tricks – this is a fact. The more you have, the fewer your opponents do – fact again. This means you can afford to bid higher when you have a good trump fit. How high can you go (assuming your high cards do not justify bidding higher)? The usual rule is that with an eight-card fit you can contract for eight tricks (a two-level contract); with a nine-card fit, you contract for nine tricks and so on.

♠ K 9 4
♥ Q J 6 2
♦ 7 3
♣ J 8 6 3

If an opponent opens 1♣ or 1♦ and partner overcalls 1♠, you should raise to 2♠ – you can envisage eight trumps. If, instead, partner overcalls 1♥, you can see a nine-card fit and so make a jump raise to 3♥ (this is not a strong bid).

♠ K 9 6 4 2
♥ A 6
♦ K J 8 7
♣ J 6

If you open 1♠, the next hand overcalls 2♣, partner
bids 2♠ and the next hand bids 3♣, you should
pass. Since partner may have only three spades,
you may well have only eight spades between you.
Partner, if holding four spades, should be the one to
compete to 3♠.

Preparing a promotion

Leading a suit in which both your partner and declarer is void can be a way of promoting a trump trick. However, declarer will be alive to the danger and may prefer to throw a loser rather than ruffing high. You counter this by making sure declarer has no losers to throw, by cashing your other winners first. Let us see this in action:

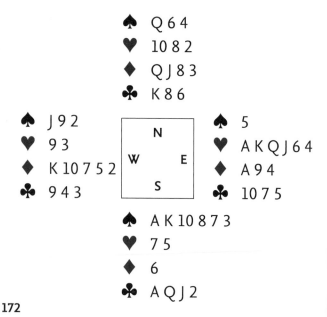

♠ Q 6 4
♥ 10 8 2
♦ Q J 8 3
♣ K 8 6

♠ J 9 2
♥ 9 3
♦ K 10 7 5 2
♣ 9 4 3

N
W   E
S

♠ 5
♥ A K Q J 6 4
♦ A 9 4
♣ 10 7 5

♠ A K 10 8 7 3
♥ 7 5
♦ 6
♣ A Q J 2

South plays in 4♠ after East has bid hearts. West leads the ♥9. If East simply cashes two top hearts and plays a third, declarer should spot West's high-low in hearts and discard a diamond – it is a loser anyway.

Before playing the third heart, East must cash the ♦A. Then declarer has to ruff the third heart. This way West will come to a trump trick however high declarer ruffs.

Cue bids for controls

You do not want to bid a slam with two fast losers (ace-king) in a suit, do you? One way to identify whether you and your partner can control a suit is for you to bid the suit if you have a control (ace, void or, by agreement, king or singleton). Cue bids to show controls occur when you have agreed a suit (directly or by implication) and the bidding is such that you cannot want to play in 3NT.

West	North	East	South
		1♦	Pass
2♠	Pass	3♠	Pass
?			

Hand 1
♠ K Q J 10 6 4
♥ 6 4 2
♦ A
♣ A Q 5

Hand 2
♠ K Q J 10 6 4
♥ A Q
♦ A Q 3
♣ 7 2

With Hand 1, you cue bid 4♣. You show a control in clubs. You need help from partner in hearts before contemplating a slam. You cannot bid 4NT because news of one ace opposite would not tell you whether there are two (or three) fast heart losers.

With Hand 2, you cue bid 4♦. The rule is you cue bid the cheaper suit first if you can. By bypassing 4♣, you thus focus partner's attention on the club suit.

Combine your chances

It is often a good idea to combine your chances rather than place all your hope on a single chance. The theme applies to both sides, though it is easier to recognize as declarer.

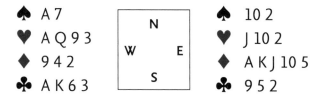

♠ A 7
♥ A Q 9 3
♦ 9 4 2
♣ A K 6 3

♠ 10 2
♥ J 10 2
♦ A K J 10 5
♣ 9 5 2

As West, you play in 3NT. North leads a spade. With six top tricks, a successful finesse in either red suit would see you home. However, a losing finesse would surely lead to immediate defeat. You should combine your chances by cashing the ace-king of diamonds. If the queen does not drop, you fall back on the heart finesse.

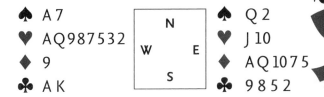

♠ A 7
♥ AQ987532
♦ 9
♣ A K

N
W   E
S

♠ Q 2
♥ J 10
♦ AQ1075
♣ 9852

As West, you play in 6♥. North leads a spade. You try the queen but South produces the king. Having won this, you should play the trump ace – perhaps the king will fall. If not, you can fall back on the diamond finesse to dispose of your spade loser. This is a better play than relying on a finesse alone in one of the red suits.

Give partner 8 points if the opponents pre-empt

If the opponents open a weak bid, at the three, four or even two level, your first action is likely to be at a higher level than you would like. To help you get into the auction, the rule is to assume that partner has a moderate 8 points.

♠ A Q
♥ K 9 5 2
♦ A 10 3
♣ K J 8 4

If the hand on your right opens three of any suit, you can overcall 3NT. Your 17 points plus partner's assumed 8 gives 25.

Hand 1
- ♠ A Q 10 6 3
- ♥ K 10 3
- ♦ K J 6 3
- ♣ 6

Hand 2
- ♠ A Q 10 6 3 2
- ♥ K 10
- ♦ A K 6 3
- ♣ 6

On Hand 1, you are happy to overcall 3♠ if the opponents open three of some other suit. Hand 2 is rather better; you do not want to play this in a part-score if partner has in the region of 8 points, so you overcall 4♠.

When partner comes in over a pre-empt, remember that 8 points for your hand is merely the expected amount. You need more than that to bid higher.

Six cards for a two-level overcall

At the one level, a five-card suit is fine for an overcall. At the two level, you normally wait for a six-card suit. Why do you need an extra card? Firstly, you are bidding at a higher level, which means an extra trick to make and a higher chance that the opponents will choose to double you. Secondly, it will often be the case that your suit is a minor and that their suit is a major. This reduces the chance that you will buy the contract (they have the higher suit) and reduces the chance you can make game (you will need 11 tricks in a minor-suit game). It also reduces the chance that partner will be on lead (if the opening is a major, responder will often raise – whereas if the opening were a minor, responder will often declare in a major or no-trumps).

♠ K 10 3
♥ K 7 3
♦ K J 9 6 4
♣ Q 4

If the hand on your right opens 1♣, a 1♦ overcall is fine. If, however, the opening is 1♥ or 1♠, you should pass. Your hand is too flat for a two-level overcall.

There are certain exceptions to the rule, of course. If you have a very strong five-card suit, perhaps A-Q-J-10-x, a two-level overcall is better than a pass or double. If your five-card suit is hearts (and they have opened 1♠), you do not want to stay out of the auction if game is on your way.

Keep trump control

When you are declarer, you do not want to run out of trumps. If you do, it is like playing in a no-trump contract. If you wanted to do that, you would have played in one.

♠ A K J 9 6 2
♥ Void
♦ Q 7 5
♣ K J 9 5

♠ 10 4
♥ 10 8 5 2
♦ A K J
♣ Q 10 6 2

As West, you play in 4♠. North leads a heart, which you ruff. If you go to dummy with a diamond and take a losing spade finesse, North will be able to make you ruff again. If trumps are 4-1, you will have none left after drawing trumps – the defenders will run the hearts after they get in with the ♣A. The answer is just to bash down the ♠A-K, accepting the possible loss of two trumps and a club.

♠ A K J 9 6
♥ J 3
♦ Q 7 5
♣ K J 9

N
W   E
S

♠ Q 10 4
♥ 10 8 5
♦ A K J
♣ Q 10 6 3

As West, you play in 4♠. You ruff the third heart and cash two trumps, finding them 4-1. If you draw the last two trumps, you will go down if the defender with the ♣A has any hearts left. Instead, you should knock out the ♣A while dummy still has a trump to deal with the fourth round of hearts.

Attack the danger hand first

If you are declarer and need to lose the lead twice, the order in which you lose the lead can be critical. If the opponents are going to set up one or more winners, you want to attack the entry to the opponent with those winners before they are ready to cash.

♠ A Q
♥ A 6 4
♦ Q J 10 8
♣ K Q J 8

♠ K 6
♥ K 7 3
♦ A 9 7 3
♣ 10 7 6 3

You are West in 3NT and receive a spade lead.

Let us suppose firstly that South opened 3♠. You place North with a doubleton spade and South with seven. Since you will go down if South has the ♦K and ♣A as entries, you assume that North has the ♣A (if the ♦K is onside, you are safe whichever minor you play first). In this case, you should take the diamond finesse (attacking the ♦K entry in the dangerous South hand) before knocking out the ♣A. If the finesse loses and South clears the spades, you assume North will have no spades left.

Now suppose that North opened 3♠. The danger hand is now North rather than South. This means you should knock out the ♣A before taking the diamond finesse.

Lead safely against 6NT

When the opponents bid up to 6NT, they are likely to have a lot of high cards and no obvious weakness. You will want to conserve the few high cards you have, hoping that the opponents take losing finesses into them. Unless the opponents have shown a long suit or you have an obvious attacking lead (e.g. K-Q in one suit and an ace elsewhere), you should lead passively.

What is a safe suit to lead? Assuming, of course, you do not have a sequence somewhere, a long weak suit is the safest to lead. A suit in which the opponents are solid is another safe option. Beware, however, of leading from two or three low in a suit the opponents have bid only modestly – your partner might have the queen and not thank you for saving declarer a two-way finesse guess.

♠ K 10 7 5 3
♥ 9 8 6
♦ Q 5 3
♣ 4 3

If the opponents bid to 3NT, you would probably lead your fourth-highest spade. Against 6NT, it is unlikely that your partner has the ace or queen of spades. The safest lead is a heart (the nine) – the opponents are less likely to be long in hearts than in clubs – a club could be horrible if partner has four to the jack for instance.

Length links to shortage

Players always start with 13 cards. If they are long in one suit, they will be short in other suits. You can use this information either as declarer or as a defender. Let us look at it from declarer's angle:

♠ K Q 10 5   ♠ A 9 4 3
♥ 8 6 4   ♥ 9 7 2
♦ A K   ♦ 9 8
♣ K J 10 4   ♣ A Q 8 6

As West you are in 4♠. The defenders take three rounds of hearts before switching to a diamond. Since all will be plain sailing if trumps are 3-2, your task is to work out which defender is more likely to hold ♠J-x-x-x. You can then cash the right honours to get a marked finesse.

Let us suppose firstly that all followed to the hearts and that South opened 3♦. In this case, South, who seems to have ten cards in the red suits, cannot possibly hold four spades. Even if South showed out on the third heart, the odds would still favour playing

North for spade length. South's diamond length suggests a shortage in spades.

Now suppose that North has bid hearts and that South turns up with a singleton heart. In this case, North has long hearts and so is the one to play for short spades. If South had a doubleton heart (and has not bid), again you would play North for the short spades.

Length links to strength

Suppose you have a two-way finesse in a suit, A-10-9-x facing K-J-x-x. Assuming you have to tackle the suit yourself, you would like to know who holds the queen.

If the opponents have bid, you may be able to tell from counting points who holds the queen. Now suppose you do not have any clues from the bidding. There could still be a way to improve on 50-50 odds. If you can find out which defender has greater length in the suit, you can finesse that defender for the queen. Whoever has greater length is more likely to hold the queen. On a 3-2 break, the queen will be in the longer holding 60% of the time; on a 4-1 break, the queen will be in the longer holding 80% of the time.

♠ K Q 10　　　　　　　♠ J 9 4
♥ A Q 8　　　　　　　♥ K J 6
♦ A 10 9 2　　　　　　♦ K J 8 3
♣ K Q 7　　　　　　　♣ A J 3

You are West in 6NT. North leads a club. Your plan should be to knock out the ♠A and cash all the winners in three suits, leaving diamonds until last. If all follow to three rounds of each suit, you might not have a lot to go on. However, often one of the suits will break 5-2, which will give you a good picture of the distribution. Someone who has five cards in one suit and three cards in two other suits can have at most two diamonds. You will play the other defender for the queen of diamonds.

Make the opponents lead

Often the best way to approach a suit is to let the other side make the first move. Of course, your opponents may be able to recognize the position too, so you need to make it hard for them to lead other suits.

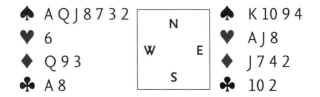

♠ A Q J 8 7 3 2
♥ 6
♦ Q 9 3
♣ A 8

♠ K 10 9 4
♥ A J 8
♦ J 7 4 2
♣ 10 2

You are West in 4♠. North leads a low heart. After you have drawn trumps, you could play diamonds yourself, with the main chance being to find South with the ten. Much better is to make the opposition play the suit. You should take the ♥A, ruff a heart, return to dummy with a trump and ruff the last heart. Then draw another round of trumps if necessary before playing the ace and another club. Whoever wins must either play a diamond or give you a ruff and discard. You are sure of success.

Other good suits to make the opponents lead are frozen suits (e.g. A-10-x facing K-9-x) and suits with a two-way finesse (e.g. K-9-x facing Q-10-x). Suits in which you have a finesse (e.g. A-Q facing x-x) are also good ones to make the opponents lead so long as you can arrange for the lead to be into rather than through the tenace.

See the unseen hands

During the play, you start by seeing only 26 cards, yours and those in dummy. From the bidding, you will often know a bit about the other hands. This will always be the case when you are a defender – declarer will always have bid. Often it is the case for declarer too. As the tricks progress, you gain more and more information.

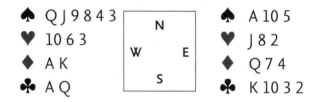

♠ Q J 9 8 4 3
♥ 10 6 3
♦ A K
♣ A Q

♠ A 10 5
♥ J 8 2
♦ Q 7 4
♣ K 10 3 2

As West, you open 1♠ in fourth seat and finish in 4♠. North cashes the ♥A-K-Q before switching to the ♦J. Do you see why you should play to drop the singleton ♠K? North, who passed as dealer, has already turned up with 10 points. South must have the ♠K.

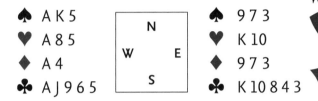

♠ A K 5		♠ 9 7 3	
♥ A 8 5	N	♥ K 10	
♦ A 4	W E	♦ 9 7 3	
♣ A J 9 6 5	S	♣ K 10 8 4 3	

As West, you open 2NT and soon arrive in 3NT.
North leads the ♠2. Since you have ten easy tricks
if clubs are 2-1, you wonder which defender might
be void. The clue here is that if North's best suit is
a four-card spade suit (the ♠2 led remember), only
South can be void in clubs.

Play for split aces

When the opponents have not bid and you have to guess who holds a missing ace, you should usually play for split aces. In other words, if one defender has already shown up with an ace, you play the other defender for the key ace.

There are two logical reasons for the rule. The first is to do with the fact that people do not lead from suits with an unsupported ace. This means the opening leader's partner will often turn up with the ace in the suit played at trick one, while the very fact that the opening leader chose not to lead some other suit could be because of holding the ace in it. The second reason is that if one defender holds two aces, further strength on the side would have led to a bid.

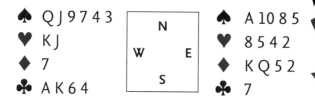

♠ Q J 9 7 4 3
♥ K J
♦ 7
♣ A K 6 4

N
W   E
S

♠ A 10 8 5
♥ 8 5 4 2
♦ K Q 5 2
♣ 7

As West you open 1♠ in fourth seat and play in 4♠. North leads the ♦4. South wins with the ♦A and switches to a low heart.

You should play for split aces, inserting the jack. Logic tells you the same thing. With the ♠K, ♥A, ♦A and any of the ♦J, ♣J or ♣Q, South would have opened.

Knowing who to play for a missing ace also helps with suits like K-Q-x-x facing J-x-x-x.

Save winners from ruffs

As declarer, you do not want the defenders to score
cheap ruffs. What is worse is when they ruff your
winners. If they ruff only your losers, their ruffs may
cause you less or no damage at all.

As West, you play in 4♠. The lead is a heart.

Let us first suppose that South opened a weak 2♥.
In this case, you read the lead as a singleton. You
should play low from dummy and win with the ace
in hand. If South gets in with ♠A and returns a heart,
North can score a ruff, yes, but the king lives to fight
another day.

Now suppose that North opened a weak 2♥ (or
overcalled in hearts). In this case, you place South
with the singleton. You should go up with dummy's

king. If North gets in with the ♠A and continues hearts, South may score a ruff but you retain your ace to win later.

If an opponent has already shown out of a suit, the position is easier. In this case you lead towards your honours, making that opponent play in the second seat.

Push only one level

Bidding can be easy when the opponents meekly pass throughout, allowing you and your partner to describe your hands. However, life is not always like that. What should you do when you were ready to bid 1♥ but someone bids 1♠ before you? What do you do if the opposing bid is 2♠ rather than 1♠?

The usual rule is that you should allow the opponents to push you one level higher than you were planning to go – but not two levels.

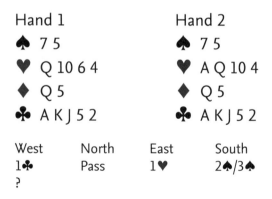

Hand 1
♠ 7 5
♥ Q 10 6 4
♦ Q 5
♣ A K J 5 2

Hand 2
♠ 7 5
♥ A Q 10 4
♦ Q 5
♣ A K J 5 2

West	North	East	South
1♣	Pass	1♥	2♠/3♠
?			

With Hand 1, you were planning to raise partner to 2♥. If the overcall is 2♠, you can raise to 3♥ instead, pushing one level. You should pass, however, if the overcall is 3♠.

With Hand 2, you were planning to rebid 3♥. This means that after a 3♠ overcall you can bid 4♥. If the overcall is 2♠, you still bid 4♥ – you would bid 3♥ with Hand 1.

Run a long suit

When defending, you do not like having to make a series of discards, do you? So, as declarer, it can be good to run a long suit. The defenders may keep the wrong things. Sometimes they cannot help but throw a crucial card.

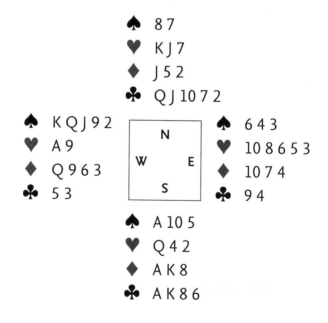

```
              ♠  8 7
              ♥  K J 7
              ♦  J 5 2
              ♣  Q J 10 7 2

♠  K Q J 9 2          ┌─────────┐          ♠  6 4 3
♥  A 9                │    N    │          ♥  10 8 6 5 3
♦  Q 9 6 3            │  W   E  │          ♦  10 7 4
♣  5 3                │    S    │          ♣  9 4
                      └─────────┘
              ♠  A 10 5
              ♥  Q 4 2
              ♦  A K 8
              ♣  A K 8 6
```

As South, you arrive in 3NT after West opened 1♠.
West leads the ♠K. You hold up the ♠A until the
third round, though you know you cannot afford to
play on hearts because West's opening bid surely
includes the ♥A.

You should run five rounds of clubs. Poor West can
spare a heart and a diamond but then has no safe
discard. A spade would let you knock out the ♥A;
a diamond would unguard the queen; the ♥A is
obviously fatal.

Set up side suits early

If you have a two-suited hand, it is often right to set up the long suit before doing other things, such as drawing trumps.

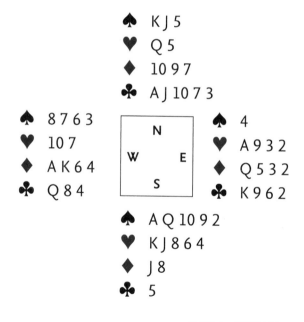

♠ K J 5
♥ Q 5
♦ 10 9 7
♣ A J 10 7 3

♠ 8 7 6 3
♥ 10 7
♦ A K 6 4
♣ Q 8 4

N
W E
S

♠ 4
♥ A 9 3 2
♦ Q 5 3 2
♣ K 9 6 2

♠ A Q 10 9 2
♥ K J 8 6 4
♦ J 8
♣ 5

As South, you arrive in 4♠. The defenders start with three rounds of diamonds. What do you do after ruffing?

If you begin by drawing trumps, you will need favourable breaks in both majors. If hearts are not 3-3, you will have a second heart loser. If spades are 4-1 and someone holds up the ♥A, you will have no way to reach the long hearts.

You should lead a low heart at trick four, playing to set up the suit. Dummy can then ruff the third round of hearts and deal with a possible fourth round of diamonds.

The five level belongs to the opponents

In a strongly competitive auction, it is usually right to sell out (pass or double) once the opponents bid to the five level. There are two good reasons for this.

Firstly, for both sides to be making, or close to making, contracts at the five level, there would have to be some extreme distribution. If one side has an eleven-card fit and the other a ten- or eleven-card fit, perhaps both sides are around the eleven-trick mark. Most of the time, bidding five over five will simply turn a plus score into a minus.

Secondly, when the opposing contract is five of a major, they are on a bit of a hiding to nothing. If they make ten tricks, they are going down. If they make twelve tricks, they are missing a slam. Even if they make exactly eleven tricks, they are no better off than if they had secured the contract a level lower.

In the case where your side is sacrificing, you do not double the opponents at the five level. If they are going down, you are probably getting a good result anyway. What you do not want to do is to turn an average score into a bad one if they make eleven tricks doubled.

When the opponents are sacrificing, obviously you do not let them play undoubled. You need to double so that the penalty you collect is close to – or ideally beats – the value of your own best contract.

Use 4NT to check on aces

To make a small slam, you need to make 12 tricks, which means losing only one. To make a grand slam, you need to make 13 tricks: this means not losing any at all. Barring voids, you cannot afford to have two aces missing in a small slam or one ace missing in a grand slam.

Once you have established that you have the playing strength for a slam and that you do not have two fast losers anywhere, the final thing to do before bidding a slam is to check on aces. The usual way to do this is to bid 4NT (the Blackwood convention). The simplest set of responses is as follows:

5♣	0 or 4 aces
5♦	1 ace
5♥	2 aces
5♠	3 aces

Since the king of trumps is almost as valuable as an ace, some people prefer to play five-ace or Keycard Blackwood, in which you count the king of trumps as an ace. The 5♦ reply then shows 1 or 5; other responses are as above.

Note that it is not such a good idea to use 4♣ to check on aces. Although doing so keeps the bidding lower, there are too many other uses for a 4♣ bid to justify sacrificing it as an ace enquiry.

Restricted choice

When an opponent drops an honour, this is more likely to be a forced play from a very short suit than a chosen play from equals. In other words, your opponent has a restricted choice.

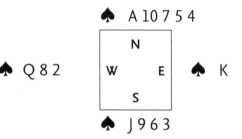

♠ A 10 7 5 4

♠ Q 8 2   N
             W   E   ♠ K
                S

♠ J 9 6 3

You (South) run the jack, losing to the king. It is not equally likely that East has the singleton king or king-queen doubleton. Half the time when holding king-queen doubleton East would have won with the queen rather than with the king. This makes it correct to finesse on the second round rather than play for the drop.

Another way of looking at this is to say that East is twice as likely to start with a singleton honour (two holdings) than with a king-queen doubleton (one holding). Nothing on the play to the first round changes this because East has to play an honour with any of the three holdings.

You might encounter many similar situations. Often you are cashing top honours rather than starting with a finesse. Such is the case on the following layouts:

```
           ♠ A K 10 7 5
          ┌───────────┐
          │     N     │
♠ Q 8 2   │ W       E │   ♠ J
          │     S     │
          └───────────┘
           ♠ 9 6 4 3
```

Having cashed the ace and seen the queen or jack fall, you should come over to the South hand and finesse West for the remaining honour. East is roughly twice as likely to hold a singleton honour as queen-jack doubleton.

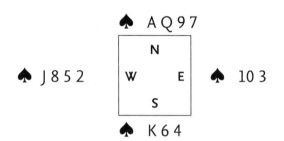

Having cashed the ace-king and seen the ten or jack fall, you should come over to the South hand and finesse West for the remaining honour. East is roughly twice as likely to hold a doubleton honour as jack-ten and a low card.

Worth a read and a watch

Reading about bridge is a great way to improve your game. When I was a teenager, I read every book in my local library and requested them from all over the county. There are so many classic books that it is hard to single out a few. My favourites included *Winning Declarer Play* by Dorothy Hayden Truscott, *Play These Hands with Me* by Terence Reese, *Why You Lose at Bridge* by S J Simon, and Victor Mollo's *Winning Double*.

The two main bridge publishers, Orion/Cassell 'Master Bridge Series' and Master Point Press both categorize their books into beginner/newcomer, intermediate/improver and advancer. These will give you an idea of the skill level assumed for the reader. Books by Ron Klinger, Eddie Kantar, and David Bird are usually safe buys. My books in the Golden Rules series you may find especially helpful.

If you play the Acol bidding system (four-card majors and a weak 1NT opening), bear in mind that books by non-British authors (other than Klinger, who does use Acol) are likely to assume a different bidding system.

If you want to watch bridge, you can do so at any time of the day or night at **www.bridgebase.com**. This site broadcasts major events all over the world with expert commentary. Even when there is no event on, you can watch players in action in online games. If a major event is taking place near to you, you can watch live at the venue.

Taking it further

Most people begin playing bridge with their friends or family. However, if you want to be competitive or you do not know many people who play bridge at the same sort of standard as yourself, you will want to widen your horizons. There are over 500 bridge clubs in the UK, which means you are unlikely to be far away from one. Bridge is a social game and joining a club is likely to improve your circle of friends as well as improving your game. If you get very keen, there are tournaments to play in at weekends. If you join enough clubs, you can play every night of the week – believe me, some people do play every day! Bridge clubs exist all over the world of course. If you want to find a club in your country, start by going to **www.ecatsbridge.com**. There you can find the website for your National Bridge Organization, from which you can then find a club near you.

If you do not have enough time to give up a whole evening, or if you live a long way from a club, you can play bridge online from the comfort of your own home. Again, the best place to start is **www.bridgebase.com**. The site offers a variety of its own clubs, including one for Acol players and a Relaxed Bridge club for players who do not take the game too seriously. You can find a partner there – or if you prefer you can hire a robot – guaranteed never to criticize your play or leave the table mid-hand!

Glossary

balanced: lacking a singleton, a void, or two doubletons

blockage: a situation in which a high card in a short suit stops you from running the suit

call: any bid, double, redouble, or pass (no bid)

convention: a bid not meaning the denomination bid

cue bid: a bid in an opponent's suit; also a bid to denote a specific holding, e.g. 1♠-(2♦)-3♦ or 1NT-3♠-4♦

declarer: the player who controls two hands

denomination: any of the four suits and no-trumps

discard: a card neither of the suit led nor a trump

doubleton: a holding of precisely two cards in a suit

duck: to play low when you have a higher card

dummy: the player with exposed cards that declarer plays

equals: cards of equal value, such as a queen-jack holding

finesse: to try to win with a card that is almost the highest

first-round control: an ace or void

fit: partnership length in a suit adequate for it to be trumps

forcing bid: a bid that asks partner to continue bidding

game-forcing: requiring that the partnership bids to game

game try: a bid that invites partner to bid game

hold up: to withhold a stopper

honour: a ten, jack, queen, king, or ace

interior sequence: a holding in which the second and third highest cards are touching but the first is not, e.g. Q-10-9

216

invitational bid: a bid that gives partner the option of passing or raising

jump: a bid to a higher level than necessary

lead: the first play to a trick

limit bid: a bid with a narrow range, e.g. 1NT-2NT

negative: a conventional response to deny specific values

opening bid: the first positive bid in an auction

opening bidder: the player who makes the first such bid

opening lead: the lead to the first trick

opening leader: the player who makes such a lead

overcall: a bid after an opponent opens, e.g. (1♦)-1♠

overruff: to ruff higher after someone else has ruffed

passed hand: a player who passed before anyone opened

penalty double: a double that you expect partner to pass

pre-empt: an obstructive jump bid with a weak hand

plain suit: in a trump contract, any of the other three suits

raise: to bid the denomination that partner has just bid

rebid: a second or subsequent bid, e.g. 1♠-2♦-2♥

response: a bid made by the partner of the opening bidder

responder: the partner of the opening bidder

reverse: to bid a lower suit and then a higher suit while going past two of the first suit, e.g. 1♣-1♥-2♦

ruff: the play of a trump on the lead of a plain suit

running suit: a suit that has all winners

second-round control: a king or singleton

sequence: two or three touching or nearly touching cards

signal: to play a card to give partner a predefined message

sign-off: a bid partner should pass, e.g. 1♥-4NT-5♣-5♥

singleton: a holding of exactly one card in a suit

solid suit: a long suit with no losers, e.g. A-K-Q-J-10-x

spot card: a card ranking below the jack

stopper: a winner in an opponent's suit

support: length in partner's suit that gives you a fit

switch or **shift:** a lead of a new suit

take-out double: a double to ask partner to bid their preferred suit, e.g. 1♥-double

tenace: two high cards of almost equal rank, e.g. A-Q

touching: of the rank just above or just below

trump: a card in the trump suit, or to ruff

unbalanced: having a void, a singleton, or two doubletons

unblock: to play a high card to avoid a blockage

underlead: to lead low from an ace or a sequence

uppercut: a high ruff to promote a trump trick for partner

void: a holding of no cards in a suit

Index

Collins

LITTLE BOOKS

These beautifully presented Little Books make excellent pocket-sized guides, packed with hints and tips.

Bananagrams Secrets
978-0-00-825046-1
£6.99

101 ways to win at Scrabble
978-0-00-758914-2
£6.99

Gin
978-0-00-825810-8
£6.99

Whisky
978-0-00-825108-6
£6.99